T0190388

Deep Belief Nets in C++ and CUDA C: Volume 1

Restricted Boltzmann Machines and Supervised Feedforward Networks

Timothy Masters

APress®

Deep Belief Nets in C++ and CUDA C: Volume 1: Restricted Boltzmann Machines and Supervised Feedforward Networks

Timothy Masters
Ithaca, New York, USA

ISBN-13 (pbk): 978-1-4842-3590-4 ISBN-13 (electronic): 978-1-4842-3591-1
https://doi.org/10.1007/978-1-4842-3591-1

Library of Congress Control Number: 2018940161

Managing Director, Apress Media LLC: Welmoed Spahr
Acquisitions Editor: Steve Anglin
Development Editor: Matthew Moodie
Coordinating Editor: Mark Powers

Cover designed by eStudioCalamar

Cover image designed by Freepik (www.freepik.com)

Distributed to the book trade worldwide by Springer Science+Business Media New York, 233 Spring Street, 6th Floor, New York, NY 10013. Phone 1-800-SPRINGER, fax (201) 348-4505, e-mail orders-ny@springer-sbm.com, or visit www.springeronline.com. Apress Media, LLC is a California LLC and the sole member (owner) is Springer Science + Business Media Finance Inc (SSBM Finance Inc). SSBM Finance Inc is a **Delaware** corporation.

For information on translations, please e-mail editorial@apress.com; for reprint, paperback, or audio rights, please email bookpermissions@springernature.com.

Apress titles may be purchased in bulk for academic, corporate, or promotional use. eBook versions and licenses are also available for most titles. For more information, reference our Print and eBook Bulk Sales web page at www.apress.com/bulk-sales.

Any source code or other supplementary material referenced by the author in this book is available to readers on GitHub via the book's product page, located at www.apress.com/9781484235904. For more detailed information, please visit www.apress.com/source-code.

Printed on acid-free paper

Table of Contents

About the Author

Timothy Masters received a PhD in mathematical statistics with a specialization in numerical computing. Since then he has continuously worked as an independent consultant for government and industry. His early research involved automated feature detection in high-altitude photographs while he developed applications for flood and drought prediction, detection of hidden missile silos, and identification of threatening military vehicles. Later he worked with medical researchers in the development of computer algorithms for distinguishing between benign and malignant cells in needle biopsies. For the last 20 years he has focused primarily on methods for evaluating automated financial market trading systems. He has authored five books on practical applications of predictive modeling: *Testing and Tuning Market Trading Systems* (CreateSpace, 2018); *Assessing and Improving Prediction and Classification* (Apress, 2018); *Data Mining Algorithms in C++* (Apress, 2018); *Neural, Novel, and Hybrid Algorithms for Time Series Prediction* (Wiley, 1995); *Advanced Algorithms for Neural Networks* (Wiley, 1995); *Signal and Image Processing with Neural Networks* (Wiley, 1994); and *Practical Neural Network Recipes in C++* (Academic Press, 1993).

About the Technical Reviewer

 **Chinmaya Patnayak** is an embedded software developer at NVIDIA and is skilled in C++, CUDA, deep learning, Linux, and file systems. He has been a speaker and instructor for deep learning at various major technology events across India. Chinmaya has an MS degree in physics and a BE degree in electrical and electronics engineering from BITS Pilani. He previously worked with Defense Research and Development Organization (DRDO) on encryption algorithms for video streams. His current interest lies in neural networks for image segmentation and applications in biomedical research and self-driving cars. Find more about him at http://chinmayapatnayak.github.io.

CHAPTER 1

Introduction

This book is intended primarily for readers who already have at least a basic knowledge of neural networks but are interested in learning about, experimenting with, and perhaps even programming deep belief nets. The salient features of this book are the following:

- The book provides motivation for the deep belief net paradigm.

- It presents the most important equations for the most common deep belief net components and justifies them to a modest degree.

- The book provides training, execution, and analysis algorithms for common deep belief net paradigms in language-independent forms.

- This book serves as a detailed user's manual for the DEEP program, which is available as a free download from the author's web site. I describe the internal operations of the program in depth.

- The book provides C++ code for many essential deep belief net algorithms. This includes versions for multiple-thread execution on Windows-based computers, as well as CUDA C implementations for using the supercomputer capabilities of NVIDIA CUDA-capable GPU cards.

It must be noted that several items are *not* included in this book.

- I largely avoid detailed mathematical theory. If we want to understand the quite advanced theory behind deep belief nets, numerous papers are available on the Internet. I will identify a few of the best later in this chapter.

- I present only those models I found to be of greatest practical, real-world value in my own work. This does not imply that the omitted models are inferior, only that I have not found them to be outstandingly useful in my particular applications.

1

T. Masters, *Deep Belief Nets in C++ and CUDA C: Volume 1*, https://doi.org/10.1007/978-1-4842-3591-1_1

In summary, I have attempted to fill gaps in the public domain material on deep belief nets. Rigorous theory is available in numerous papers, especially those of Dr. Geoffrey Hinton and other pioneers in the field. Reproducing these excellent discussions would be redundant. Also, general statements of basic algorithms are widely available on the Internet, though these are generally devoid of the practical nuances that make the difference between a useful algorithm and something suitable for only toy problems. What appears to be lacking in the public domain are the specific, practical bits of information needed by someone who wants to program deep belief nets and use them to solve real-world problems. This book focuses on such practicalities.

Review of Multiple-Layer Feedforward Networks

A multiple-layer feedforward network is generally illustrated as a stack of layers of "neurons," similar to what is shown in Figures 1-1 and 1-2. The bottom layer is the input to the network, what would be referred to as the *independent variables* or *predictors* in traditional modeling literature. The layer above the input layer is the *first hidden layer*. Each neuron in this layer attains an *activation* that is computed by taking a weighted sum of the inputs and then applying a nonlinear function. Each hidden neuron in this layer will have a different set of input weights.

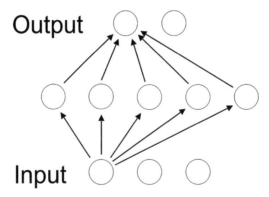

Figure 1-1. *A shallow network*

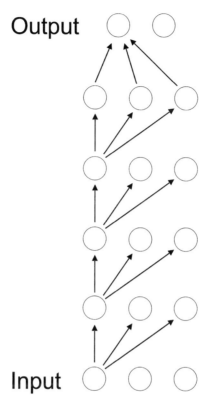

Figure 1-2. *A deep network*

If there is a second hidden layer, the activations of each of its neurons are computed by taking a weighted sum of the activations of the first hidden layer and applying a nonlinear function. This process is repeated for as many hidden layers as desired.

The topmost layer is the output of the network. There are many ways of computing the activations of the output layer, and several of them will be discussed later. For now let's assume that the activation of each output neuron is just a weighted sum of the activations of the neurons in the prior layer, without use of a nonlinear function.

Figures 1-1 and 1-2 show only a small subset of the connections. Actually, every neuron in every layer feeds into every neuron in the next layer above.

To be more specific, Equation 1-1 shows the activation of a hidden neuron, expressed as a function of the activations of the prior layer. In this equation, $x = \{x_1, ..., x_K\}$ is the vector of prior-layer activations, $w = \{w_1, ..., w_K\}$ is the vector of associated weights, and b is a bias term.

$$a = f\left(b + \sum_{k=1}^{K} w_k x_k\right) \qquad (1\text{-}1)$$

It's often more convenient to consider the activation of an entire layer at once. In Equation 1-2, the weight matrix W has K columns, one for each neuron in the prior layer, and as many rows as there are neurons in the layer being computed. The bias and layer inputs are column vectors. The nonlinear activation function is applied element-wise to the vector.

$$a = f(b + Wx) \qquad (1\text{-}2)$$

There is one more way of expressing the computation of activations, which is most convenient in some situations. The bias vector b can be a nuisance, so it can be absorbed into the weight matrix W by appending it as one more column at the right side. We then augment the x vector by appending 1.0 to it: $x = \{x_1, ..., x_K, 1\}$. The equation for the layer's activations then simplifies to the activation function operating on a simple matrix/vector multiplication, as shown in Equation 1-3.

$$a = f(Wx) \qquad (1\text{-}3)$$

What about the activation function? Traditionally, the hyperbolic tangent function has been used because it has some properties that make training faster. However, for reasons that will become clear later, we will exclusively use the logistic function shown in Equation 1-4 and graphed in Figure 1-3.

$$f(t) = \frac{1}{1 + e^{-t}} \qquad (1\text{-}4)$$

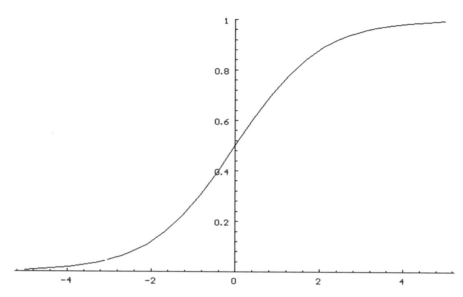

Figure 1-3. *The logistic activation function*

There are numerous theorems that show the power of a neural network having even a single hidden layer. We will not pursue these here, but know that in a broad class of problems such a network is theoretically capable of solving the problem. Adding a second hidden layer for all practical purposes mops up the few remaining issues. So, it's no surprise that multiple-layer feedforward networks are so popular.

What Are Deep Belief Nets, and Why Do We Like Them?

Prior to the development of neural networks, researchers generally relied on large doses of human intelligence when designing prediction and classification systems. One would measure variables of interest and then brainstorm ways of massaging these "raw" variables into new variables that (at least in the mind of the researcher) would make it easier for algorithms such as linear discriminant analysis to perform their job. For example, if the raw data were images expressed as arrays of gray-level pixels, one might apply edge detection algorithms or Fourier transforms to the raw image data and feed the results of these intermediate algorithms into a classifier.

The data-analysis world shook when neural networks, especially multiple-layer feedforward networks, came into being. Suddenly prediction and classification tools, compared to earlier methods, relied to a much lesser degree on human-driven preprocessing. It became feasible to simply present an array of gray-level pixels to a neural network and watch it almost miraculously discover salient class features on its own.

For many years, the prevailing wisdom stated that the best architecture for a feedforward neural network was *shallow* and wide. In other words, in addition to the input (often called the *bottom* layer) and the output (often called the *top* layer), the network would have only one, or perhaps two at most, intervening *hidden* layers. This habit was encouraged by several powerful forces. Theorems were proved showing that in broad classes of problems, one or two hidden layers were sufficient to solve the problem. Also, attempts to train networks with more than two hidden layers almost always met with failure, making the decision of how many layers to use a moot point. According to the theorems of the day, you didn't need deeper networks, and even if you did want more layers, you couldn't train them anyway, so why bother trying?

The fly in the ointment was the fact that the original selling point of neural networks was that they supposedly modeled the workings of the brain. Unfortunately, it is well known that brains are far from shallow in their innermost computational structure (except for those of a few popular media personalities, but we won't go there). Then new theoretical results began appearing that showed that for many important classes of problems, a network composed of numerous narrow layers would be more powerful than a wider, shallower network having the same number of neurons. In effect, although a shallow network might be *sufficient* to solve a problem, it would require enormous width to do so, while a deep network could solve the problem even though it may be very narrow. Deep networks proved enticing, though still enormously challenging to implement.

The big breakthrough came in 2006 when Dr. Geoffrey Hinton et al. published the landmark paper "A Fast Learning Algorithm for Deep Belief Nets." This work laid the foundation for what is, roughly stated, a good general procedure for training deep networks.

1) Use unsupervised training to teach a single layer of hidden neurons to reflectively reproduce the raw data inputs without regard to the desired outputs. If there are fewer hidden neurons than inputs, this forces the layer to learn patterns rather than rote memorization to spit back. This pattern learning resembles what happens with principal components or some types of data compression. Even if there are not fewer hidden neurons than inputs, pattern learning can occur if training is done properly.

2) For each training case, use the activations of the trained hidden layer as inputs to another hidden layer. Teach this second hidden layer to reflectively reproduce its inputs.

3) Repeat this for as many hidden layers as desired. Each new hidden layer encodes higher-level patterns. For example, the raw data input may be a rectangular array of gray-level pixels. The first hidden layer may detect edges in various parts of the image. The second hidden layer may detect patterns of connection of these edges. A third hidden layer may assemble recognizable objects out of the connected edges. Such progressive feature discovery is impossible in a shallow network.

4) Use supervised training to map the activations of the hidden neurons in the last (topmost) hidden layer to classes or predictions. Alternatively, use the weights learned in the unsupervised training as initial values for supervised training of a traditional but very deep feedforward neural network.

Many variations of this general algorithm are possible, and we will explore several of the most popular and useful variations in this book. But in a broad sense, the algorithm described here covers much of the application of deep belief nets today.

One of the most fascinating properties of deep belief nets is their remarkable ability to generalize beyond the universe of training examples. This is likely because the output layer, rather than seeing the raw data, is seeing "universal" patterns in the raw data—patterns that because of their universality are likely to reappear in the general population.

A closely related property is that deep belief nets are shockingly robust against overfitting. Every beginning statistics student learns the importance of using many, many more training cases than optimizable parameters. The standard wisdom is that if one uses 100 cases to train a model with 50 optimizable parameters, the resulting model will learn as much about the noise in the training set as it learns about the legitimate patterns and will hence be worthless. But consider the MNIST dataset of handwritten digits. Its standard training set has 60,000 training cases and 10,000 independent test cases. One can use these 60,000 cases to train a deep belief net that has several *million* optimizable parameters and yet achieve on the order of 1 percent error rate on the test set!

In all likelihood, the reason for this robustness against overfitting is the fact that until the final layer is trained, all learning is unsupervised. During training of the hidden layers, the learning algorithm knows nothing about the true class of each training case. (Actually, there are some hybrid algorithms that employ limited knowledge of the true class, and supervised fine-tuning of the hidden layers is sometimes done, but we'll put off discussion of such algorithms for now.) Because the training algorithm must operate without knowing true class membership, it must instead learn to recognize consistent patterns in the data. By definition, noise is unlikely to have consistent patterns. Only legitimate patterns will appear regularly in the data. This provides incentive for the model to learn the actual structure of the data while ignoring random noise.

CHAPTER 2

Supervised Feedforward Networks

Deep belief nets are generally trained in stages. First, one or more (usually more) layers are trained with unsupervised algorithms. Rather than seeking to learn class memberships or predicted values, the model simply tries to find consistent patterns within the independent variables. Only after such patterns have been found does training switch to supervised mode. However, because supervised training algorithms are easier to understand than the usual unsupervised algorithms, we will begin this study of deep belief nets with supervised training.

Backpropagation of Errors

The fundamental goal of supervised training can be summarized simply: find a set of parameters (weights and biases as in Equation 1-2) such that, given an input to the neural network, the output of the network is as close as possible to the desired output. To find such parameters, we must have a performance criterion that rigorously defines the concept of *close*. We then find parameters that optimize this criterion.

A traditional favorite criterion is the *mean squared error*. For each training case, we sum the squared differences between the desired activation of each output neuron and the obtained activation. The mean of squared differences, taken across all training cases, has numerous desirable theoretical and practical properties that are outside the scope of this text.

© Timothy Masters 2018
T. Masters, *Deep Belief Nets in C++ and CUDA C: Volume 1*, https://doi.org/10.1007/978-1-4842-3591-1_2

Suppose we have K output neurons numbered 1 through K. For a given training case, let t_k be the true value for this case, the value that we hope the network will produce, and let o_k be the output actually obtained. Then the mean squared error (MSE) for this single case is given by Equation 2-1. To compute the MSE for the entire training set, sum this quantity for all cases and divide by the number of cases.

$$E = \frac{1}{K}\sum_{k=1}^{K}\left(o_k - t_k\right)^2 \tag{2-1}$$

Supervised training of a multiple-layer feedforward network amounts to finding the weights and bias terms that minimize Equation 2-1. In any numerical minimization algorithm, it is of great benefit to be able to efficiently compute the gradient, the partial derivatives of the criterion being minimized with respect to each individual parameter. Luckily, this is quite easy in this application. We just start at the output layer and work backward, repeatedly invoking the chain rule of differentiation.

The activation of output neuron k is just a weighted sum of the activations of the neurons in the prior layer. We'll use the Greek letter delta to designate the derivative of the error with respect to this weighted sum coming into output neuron k. Examination of Equation 2-1 shows this to simply be as shown in Equation 2-2. In this equation, the superscript O means that this signifies the delta of an output neuron.

$$\delta_k^O = 2\left(o_k - t_k\right) \tag{2-2}$$

This neuron is receiving a weighted sum of activations from all neurons in the prior layer. How can we compute the derivative of the error with respect to the weight from neuron i in the prior layer? The simple chain rule says that this is the product of the derivative in Equation 2-2 times the derivative of the net input (the weighted sum coming into this output neuron) with respect to this weight.

This latter term is trivial. The contribution to the weighted sum from neuron i in the prior layer is just the activation of that neuron times the weight connecting it to the output neuron k. We shall designate this output weight as w_{ki}^O. So, the derivative of that weighted sum with respect to w_{ki}^O is just the activation of neuron i. This leads us to the formula for the partial derivatives of the error with respect to the weights connecting the last hidden layer to the output layer. In Equation 2-3 we use the superscript M on a to indicate that it is the activation of a neuron in hidden-layer M, where there are M hidden layers numbered from 1 through M.

$$\frac{\partial E}{\partial w_{ki}^o} = a_i^M \delta_k^o \tag{2-3}$$

There are two complications when we deal with the weights feeding hidden layers. Let's consider the weights leading from hidden-layer $M-1$ to hidden-layer M, the last hidden layer. We ultimately want the partial derivatives of the error with respect to each of these weights. As when dealing with the output layer, we'll split this derivative into the product of the derivative of the net input feeding this neuron with respect to the weight times the derivative of the error with respect to this neuron's net input. As before, the former term here is trivial; it's just the activation of the prior neuron feeding through this weight. It's the latter that's messy.

The first complication is that the outputs are linear: the output activation is just the net input to that output neuron. But the hidden neurons are nonlinear. In particular, the function that maps the net input of a hidden neuron to its activation is the logistic function shown in Equation 1-4. So, the chain rule states that the derivative of the error with respect to the net input is the derivative of the error with respect to the output times the derivative of the output with respect to the input. Luckily, the derivative of the logistic function $f(a)$ is simple, as shown in Equation 2-4.

$$f'(a) = f(a)\left(1 - f(a)\right) \tag{2-4}$$

The remaining term is more complicated because the output of a neuron in a hidden layer feeds into every neuron in the next layer and thus impacts the error through every one of those paths. Recall that δ_k^o is the derivative of the error with respect to the weighted sum coming into output neuron k. The contribution to this weighted sum going into output neuron k from neuron i in the prior layer M is the activation of hidden neuron i times the weight connecting it to output neuron k. So, the impact on the derivative of the error from the activation of neuron i that *goes through this path* is δ_k^o times the connecting weight. Since neuron i impacts the error through all output neurons, we must sum these contributions, as shown in Equation 2-5.

$$\frac{\partial E}{\partial a_i^M} = \sum_{k=1}^{K} w_{ki}^o \, \delta_k^o \tag{2-5}$$

Pant pant. We are almost there. Our goal, the partial derivative of the error with respect to the weight connecting a neuron in hidden-layer $M-1$ to a neuron in hidden-layer M, is the product of the three terms which we have already presented, listed here:

- The derivative of the net input to the neuron in hidden-layer M with respect to the weight in which we are interested

- The derivative of the output of this neuron with respect to its net input (the derivative of its nonlinear activation function)

- The derivative of the error with respect to the output of this neuron

The derivative of the error with respect to w_{ij}^M (the weight connecting neuron j in layer $M-1$ to neuron i in layer M) is the product of these three terms. The product of the second and third of these terms is given by Equation 2-6, with $f'(.)$ being given by Equation 2-4. The multiplication is completed in Equation 2-7.

$$\delta_i^M = f'\left(a_i^M\right)\sum_{k=1}^{K} w_{ki}^O\, \delta_k^O \tag{2-6}$$

$$\frac{\partial E}{\partial w_{ij}^M} = a_j^{M-1}\delta_i^M \tag{2-7}$$

There is no need to derive the equations for partial derivatives of weights in hidden layers prior to the last hidden layer, as the equations are the same, just pushed back one layer at a time by successive application of the chain rule. In particular, for some hidden-layer $m<M$, we have Equation 2-8 for the partial derivative of the error with respect to the weighted sum coming into neuron i in layer m. Equation 2-9 then provides the partial derivative of the error with respect to the weight connecting neuron j in hidden-layer $m-1$ to neuron i in hidden-layer m. In this case, there are K neurons in hidden-layer $m+1$.

$$\delta_i^m = f'\left(a_i^m\right)\sum_{k=1}^{K} w_{ki}^{m+1}\, \delta_k^{m+1} \tag{2-8}$$

$$\frac{\partial E}{\partial w_{ij}^m} = a_j^{m-1}\delta_i^m \tag{2-9}$$

SoftMax Outputs for Classification

The discussion so far has been restricted to using mean squared error as the measure of quality. This is usually an excellent choice when the goal is numeric prediction. But MSE is problematic when the goal is classification. In addition to some subtle problems that are beyond the scope of this text, there are two obvious problems.

First, the usual and reasonable approach to classification is to use a separate output neuron for each class and train the model to attain an output value of one for the correct class and zero for all other classes. When a trial case is presented to the model, the class decision is whichever output has the highest value. But wouldn't it be nice if the outputs could be interpreted as probabilities? The outputs described so far can even be negative, which is not conducive to their use as probabilities!

Figure 2-1 illustrates another serious problem with using MSE as a performance criterion. This is a fantasy-easy classification problem. We have two predictor variables, *X1* and *X2*, and a single binary target, which is zero for one class and one for the other class. This figure is a scatterplot of the training cases. Obviously, we can find a linear combination of these two predictors that produces perfect class separation, as shown by the dotted line. Since we are trying to discriminate between target values of zero and one, setting a prediction threshold of 0.5 seems reasonable.

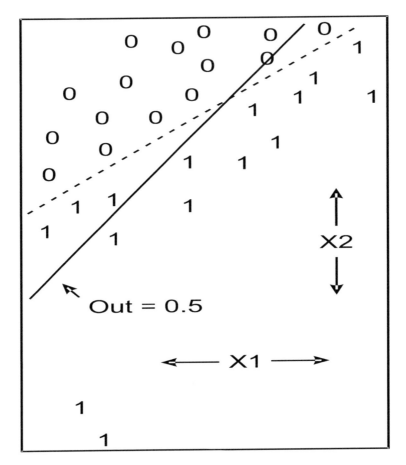

Figure 2-1. *MSE makes for poor classification*

There are two oddball cases in the lower left of the plot. These are awesome members of the 1 class. Unfortunately, a linear model corresponding to the dotted line would leave these two outliers quite distant from the truly optimal decision boundary. This ideal model would be producing an output of around 3.0 or so, which produces a large MSE when compared to the desired prediction of 1.0. When we minimize MSE, the resulting model will shift the decision boundary, which as noted earlier is perhaps the line for which the output is 0.5, in such a way that these two outliers are not so far out in left field. In effect, the model is sacrificing practical performance (classification) to improve a type of performance (MSE) that is irrelevant to the task.

An excellent solution to this problem is to make use of the long-standing and common method of handling the exact problem just discussed: logistic regression. This venerable technique uses a linear combination of variables (plus a bias term) to predict the log of the *odds ratio* of class membership. In the two-class case, let π be the probability that a case belongs to a class of interest. Then the odds ratio is $\pi/(1-\pi)$. Equation 2-10 shows the logistic regression model.

$$\log\left(\frac{\pi}{1-\pi}\right) = b + \sum_i w_i x_i \tag{2-10}$$

So, instead of asking this linear combination to predict either the exact value 1.0 or 0.0 according to class membership, it is instead being asked to predict a continuous value closely associated with a probability. When logistic regression is used for the output layer of a multiple-layer feedforward network, the x_i terms in Equation 2-10 would be the activations of the neurons in the final hidden layer.

Equation 2-10 can be generalized to the multiple-class situation easily. We will once again assume that the bias terms are absorbed into the weights by appending 1.0 to every x vector, similar to what was done in Equation 1-3. Let w_k be the vector of weights (with the bias appended) for computing the value fed to output neuron k. This value, the dot product of the weights and x, is often called the *logit* and is defined in Equation 2-11. In this context, x is the vector of activations of the final hidden layer, with 1.0 appended for the bias term.

$$logit_k = w_k \bullet x \tag{2-11}$$

Suppose there are K classes. Equation 2-12 gives the model's estimated probability that the case that produced x belongs to class k (the activation of output neuron k). These are called *SoftMax* activations because they are a smoother version of "hard max" winner-takes-all classification in which the model simply chooses the class that has maximum activation.

$$p(y=k) = \frac{e^{logit_k}}{\sum_{i=1}^{K} e^{logit_i}} \tag{2-12}$$

It should be obvious that these output activations are non-negative and sum to one. Just as in the prior situation of predicting one or zero and minimizing MSE, we have K weight vectors, each of which has $m+1$ elements, where m is the number of neurons in the last hidden layer.

But wait, astute readers will say, we are now constrained because the output activations must sum to one. The implication is that if we know $K-1$ activations, the remaining activation is determined. So, for SoftMax we need only $K-1$ weight vectors. Great observation!

What we can do is assume that the weight vector for the last class is identically zero. (Actually, we could choose any class for this distinction and the results would be the same. But tradition rules.) In this case, Equation 2-12 takes two forms according to the class. For all classes except the last, we use Equation 2-13, and for the last class we use Equation 2-14. Note that the 1.0 in the denominator comes from the fact that $e^0=1$.

$$p(y=k)=\frac{e^{logit_k}}{1+\sum_{i=1}^{K-1} e^{logit_i}} \quad if \quad k<K \tag{2-13}$$

$$p(y=K)=\frac{1}{1+\sum_{i=1}^{K-1} e^{logit_i}} \tag{2-14}$$

The mean squared error optimization criterion was defined in Equation 2-1. We now need a different optimization criterion to find good values for the parameters (weights and biases) for the SoftMax output layer. The best choice is maximum likelihood. Any set of parameters defines, by means of the equations just shown, the probability of each possible class given x. The training set is assumed to be random draws from a population, each of which provides an x vector and a true class. If we were to consider a given set of model parameters as defining the true model, we could compute (in a sense best left undiscussed here) the probability of obtaining the set of training cases that were observed. So, we find that set of parameters that maximizes this probability. In other words, we seek the model that provides the maximum likelihood of having obtained the training set in these random draws from the population.

In this particular application, the likelihood of a case is just the probability given by the model for the class to which that case belongs. We want a criterion that is summable across the training set, so instead of considering the likelihood, which is multiplicative, we will use the log likelihood as your criterion. This way we can compute the criterion for the entire training set by summing the values for the individual cases in the training set.

Also, to conform to more general forms of the log likelihood function we may encounter in more advanced texts, as well as to conform to the expression of the derivative that will soon be discussed, we express the log likelihood of a case in a more complex manner. For a given training case, define t_k as 1.0 if this case is a member of class k, and 0.0 otherwise. Also, define p_k as the activation of output neuron k, as given by Equation 2-12 or the two equivalent equations, Equation 2-13 and Equation 2-14. Then the log of the likelihood corresponding to the model's parameters is given by Equation 2-15. This equation is called the *cross entropy*, and interested readers might wish to can look up this term for some fascinating insights.

$$L = \sum_{k=1}^{K} t_k \log(p_k) \tag{2-15}$$

Observe that in the inner summation over classes, every term is zero except the term corresponding to the correct class. Thus, the log likelihood is just the log of the model's computed probability for the correct class of the case. Here are some observations about the log likelihood:

- Because p is less than one, the log likelihood is always negative.

- The better the model is at computing the correct class probabilities, the larger (closer to zero; smaller magnitude) this quantity will be since it is the log probability of the correct class, and a good model will provide a large probability for the correct class.

- If the model is perfect, meaning that the computed probability of the correct class will be 1.0 for every case, the log likelihood will be zero, its maximum possible value.

Now comes a bit of almost unbelievable luck. Recall that Equation 2-2 gave us the derivative of the MSE with respect to the weighted sum coming into output neuron k. This was trivial calculus. Just imagine how horrendously complicated will be the formula for the derivative of the log likelihood shown in Equation 2-15, especially given the complexity of the probabilities defined in Equation 2-12 and its two equivalents. But here's the surprise. Without going through a considerable number of steps, we can state that this derivative of Equation 2-15 for a case is given by Equation 2-16. Yup. It's that simple.

$$\delta_k^o = \frac{\partial L}{\partial logit_k} = p_k - t_k \tag{2-16}$$

Amazingly, except for a factor of two, the delta for a SoftMax output layer and maximum likelihood optimization is identical to that for a linear output layer and mean-squared-error optimization. All subsequent equations for the gradient hold. So, these two different approaches to modeling can be handled with almost the same code.

Code for Gradient Calculation

All of the routines discussed in this book can be downloaded free of charge from the author's web site. Nonetheless, it is instructive to list many of them in the text so that I can explain how they operate.

We begin with a workhorse routine that will be used heavily in supervised training. This routine simply computes the dot product of two vectors: the sum of their element-wise products. However, it's shown here because it illustrates an unusual approach that can provide a significant speed-up in some hardware. Here it is:

```
double dotprod (
   int n ,             // Length of vectors
   double *vec1 ,   // One of the vectors to be dotted
   double *vec2 )   // The other vector
{
   int k, m ;
   double sum ;

   sum = 0.0 ;  // Will cumulate dot product here
   k = n / 4 ;    // Divide vector into this many groups of 4
   m = n % 4 ;  // This is the remainder of that division

   while (k--) {    // Do each group of 4
     sum += *vec1 * *vec2 ;
     sum += *(vec1+1) * *(vec2+1) ;
     sum += *(vec1+2) * *(vec2+2) ;
     sum += *(vec1+3) * *(vec2+3) ;
     vec1 += 4 ;
     vec2 += 4 ;
     }
```

```
  while (m--)      // Do the remainder
    sum += *vec1++ * *vec2++ ;

  return sum ;
}
```

This subroutine implements a technique called *loop unrolling*. Many CPUs look into the future in their execution stream and try to "work ahead" as much as possible by doing things such as starting to fetch data from memory before it's needed so that it's ready when called for. But for some processors, hitting the end of a loop grinds this process to a halt because it does not know which way a branch will go: repeat the loop or pass on down. The best modern processors and compilers make effective use of branch prediction heuristics to minimize the impact of such pipeline stalls, but some don't, and in any case it can never be perfect. So, fanatics will perform multiple calculations in a row before handling loop logic. If the vectors are long and the processor heavily reliant on pipelining but not so good at branch prediction, loop unrolling can make a significant difference in execution time. It's always worth trying, and it certainly doesn't hurt.

The next subroutine is also called often. Many users will want to forego its use and instead embed this small amount of code into the calling routines. But by making this common operation a separate routine, subsequent code becomes shorter and clearer. This routine takes an input vector and computes the activation of a single neuron. Normally it would be called with outlin true for the output layer to produce a linear output, and false for hidden neurons so that the logistic activation function is used.

```
void activity (               // Implement Equation 1-1 on Page 4
   double *input ,            // This neuron's input vector, ninputs long
   double *coefs ,            // Weight vector, ninputs+1 long (bias is at end)
   double *output ,           // Achieved activation of this neuron
   int ninputs ,              // Number of inputs
   int outlin                 // Activation function is identity if nonzero, else logistic
   )
{
   double sum ;

   sum = dotprod ( ninputs , input , coefs ) ;
   sum += coefs[ninputs] ;    // Bias term
```

```
  if (outlin)
    *output = sum ;
  else
    *output = 1.0 / (1.0 + exp(-sum)) ;
}
```

Now let's step up the complexity a little. The following subroutine executes the entire network in that it takes an input vector and computes the activations of all hidden and output neurons. The _thr in its name is my own flag that this is a thread-safe version of a subroutine that also exists in a not-so-safe form.

Some of the calling parameters deserve explanation because this routine has been pulled directly from the much larger and more complex DEEP program. The suffix _all used in some names is a flag that means the variable is talking about the entire network, as opposed to just part of it. (Later we will see that the entire deep model is subdivided into sections.) The array nhid_all holds the number of neurons in each hidden layer. The array of pointers weights_opt holds pointers to the weights for each layer. The weights in layer i, pointed to by weights_opt[i], consist of nhid_all[i] sets of weights, each set consisting of as many elements as there are neurons in the prior layer (or model inputs, for the first hidden layer), plus one for the bias. The array hid_act contains pointers to the activation vector for each hidden layer, with each vector containing as many elements as there are neurons in this layer. These values will be computed and saved for possible subsequent use. Finally, final_layer_weights contains ntarg sets of weight vectors for the output layer, with each weight vector containing as many elements as there are neurons in the final hidden layer, plus one for the bias.

```
static void trial_thr (
    double *input ,           // Input vector n_model_inputs long
    int n_all ,               // Number of layers, including output, not including input
    int n_model_inputs ,      // Number of inputs to the model
    double *outputs ,         // Output vector of the model, ntarg long
    int ntarg ,               // Number of outputs
    int *nhid_all ,           // nhid_all[i] is the number of hidden neurons in hidden layer i
    double *weights_opt[] ,   // weights_opt[i] points to the weight vector for hidden layer i
    double *hid_act[] ,       // hid_act[i] points to the vector of activations of hidden layer i
    double *final_layer_weights , // Weights of final layer
    int classifier            // If nonzero use SoftMax output; else use linear output
    )
```

```
{
   int i, ilayer ;
   double sum ;

   for (ilayer=0 ; ilayer<n_all ; ilayer++) {
      if (ilayer == 0 && n_all == 1) {          // Direct input to output? (No hidden)
         for (i=0 ; i<ntarg ; i++)
            activity ( input , final_layer_weights+i*(n_model_inputs+1) ,
                     outputs+i , n_model_inputs , 1 ) ;
         }

      else if  (ilayer == 0) {                  // First hidden layer?
         for (i=0 ; i<nhid_all[ilayer] ; i++)
            activity ( input , weights_opt[ilayer]+i*(n_model_inputs+1) ,
                     hid_act[ilayer]+i , n_model_inputs , 0 ) ;
         }

      else if (ilayer < n_all-1) {              // Subsequent hidden layer?
         for (i=0 ; i<nhid_all[ilayer] ; i++)
            activity ( hid_act[ilayer-1] , weights_opt[ilayer]+i*(nhid_all[ilayer-1]+1) ,
                     hid_act[ilayer]+i , nhid_all[ilayer-1] , 0 );
         }

      else {                                    // Output layer
         for (i=0 ; i<ntarg ; i++)
            activity ( hid_act[ilayer-1] , final_layer_weights+i*(nhid_all[ilayer-1]+1) ,
                     outputs+i , nhid_all[ilayer-1] , 1 );
         }
      }

   if (classifier) { // Classifier is always SoftMax (Equation 2-12) on Page 15
      sum = 0.0 ;
      for (i=0 ; i<ntarg ; i++) {               // For all outputs
         if (outputs[i] < 300.0)                // SoftMax can occasionally produce huge outputs
            outputs[i] = exp ( outputs[i] ) ;
         else
            outputs[i] = exp ( 300.0 ) ;
         sum += outputs[i] ;
         }
```

```
    for (i=0 ; i<ntarg ; i++)
      outputs[i] /= sum ;
    }
}
```

Now we come to the real core of the code, that for computing the optimization criterion and gradient. We'll break up this subroutine into sections and discuss each separately. Here are the calling parameters:

```
double batch_gradient (
  int istart ,                    // Index of starting case in input matrix
  int istop ,                     // And one past last case
  double *input ,                 // Input matrix; each case is max_neurons long
  double *targets ,               // Target matrix; each case is ntarg long
  int n_all ,                     // Number of layers, including output, not including input
  int n_all_weights ,             // Total num of weights, including final layer and all bias terms
  int n_model_inputs ,            // N of inputs to the model; Input matrix may have more cols
  double *outputs ,               // Output vector of the model; used as work vector here
  int ntarg ,                     // Number of outputs
  int *nhid_all ,                 // nhid_all[i] is the number of hidden neurons in hidden layer i
  double *weights_opt[] ,         // weights_opt[i] points to the weight vector for hidden layer i
  double *hid_act[] ,             // hid_act[i] points to the vector of activations of hidden layer i
  int max_neurons ,               // N of columns in input matrix; max exceed n_model_inputs
  double *this_delta ,            // Delta for the current layer
  double *prior_delta ,           // And saved for use in the prior (next to be processed) layer
  double **grad_ptr ,             // grad_ptr[i] points to gradient for layer i
  double *final_layer_weights ,   // Weights of final layer
  double *grad ,                  // All computed gradients, strung out as a single long vector
  int classifier                  // If nonzero use SoftMax output; else use linear output
  )
```

Most of the parameters in the calling list have already been discussed, but a few are new. This routine is able to be called for subsets of the entire training set, so the starting and stopping (one past the last) indices are specified. The gradient, which encompasses all layers, is strung out into a single vector, grad, which contains n_all_weights elements.

The trickiest parameter is grad_ptr. This contains as many elements as there are layers in the model, including the output layer but not including the input. Each element in grad_ptr is a pointer to the beginning of the gradient vector (in grad) for the corresponding layer. Each gradient vector is a set of subvectors, one for each neuron in the layer. Each subvector contains as many elements as there are neurons in the prior layer (or inputs for the first hidden layer), plus one for the bias.

The first thing done is to initialize for the cumulation of the gradient and error (or negative log likelihood for SoftMax) across the batch of training cases. Then the loop that processes each training case in the batch begins. The training dataset may contain more columns than the network has inputs, although the inputs will always be together in the first columns. So, the pointer to the current training case, dptr, takes this into account. The entire network is activated for this case by calling trial_thr, which has already been discussed. Then we get a pointer to the target vector that corresponds to this training case. If this model is being used for prediction, the target vector may contain any real values. But if it is a classifier, in which case a SoftMax output layer will be used, then the target vector usually contains 1.0 for the correct class and 0.0 for all other classes.

```
{
  int i, j, icase, ilayer, nprev, nthis, nnext, imax ;
  double diff, *dptr, error, *targ_ptr, *prevact, *gradptr, delta, *nextcoefs, tmax ;

  for (i=0 ; i<n_all_weights ; i++)      // Zero gradient for summing
    grad[i] = 0.0 ;                       // All layers are strung together here

  error = 0.0 ;                          // Will cumulate total error here

  for (icase=istart ; icase<istop ; icase++) {

    dptr = input + icase * max_neurons ; // Point to this sample
    trial_thr ( dptr , n_all , n_model_inputs , outputs , ntarg , nhid_all ,
            weights_opt , hid_act , final_layer_weights , classifier ) ;

    targ_ptr = targets + icase * ntarg ;
```

The next step is to compute and cumulate the optimization criterion. If the output is SoftMax, we find the true class by looking for the largest target value. While doing this, we use Equation 2-16 to compute the output delta. The contribution of this case to the negative log likelihood is given by Equation 2-15. Recall that the algorithm is minimizing the error but we want to maximize the likelihood, so we flip its sign. If the output is not SoftMax, we just cumulate the MSE and compute delta with Equation 2-2.

```
if (classifier) {           // SoftMax
  tmax = -1.e30 ;
  for (i=0 ; i<ntarg ; i++) {   // Find the true class as that having max target
    if (targ_ptr[i] > tmax) {
      imax = i ;
      tmax = targ_ptr[i] ;
      }
    this_delta[i] = targ_ptr[i] - outputs[i] ; // Neg deriv of cross entropy wrt input (logit) i
    }
  error -= log ( outputs[imax] + 1.e-30 ) ; // We minimize negative log likelihood
  }

else {
  for (i=0 ; i<ntarg ; i++) {
    diff = outputs[i] - targ_ptr[i] ;
    error += diff * diff ;
    this_delta[i] = -2.0 * diff ; // Neg deriv of squared error wrt input to neuron i
    }
  }
```

Next we compute the gradient of the output layer, the (negative) partial derivatives of the criterion with respect to the weights feeding the output neurons. Here is the code, and an explanation will follow:

```
if (n_all == 1) {                    // No hidden layer
  nprev = n_model_inputs ;           // Number of inputs to the output layer
  prevact = input + icase * max_neurons ; // Point to this sample
  }
else {
  nprev = nhid_all[n_all-2] ;        // n_all-2 is the last hidden layer
  prevact = hid_act[n_all-2] ;       // Point to layer feeding the output layer
  }
```

```
gradptr = grad_ptr[n_all-1] ;      // Point to output gradient in grand gradient vector
for (i=0 ; i<ntarg ; i++) {         // For all output neurons
   delta = this_delta[i] ;          // Neg deriv of criterion wrt logit
   for (j=0 ; j<nprev ; j++)
      *gradptr++ += delta * prevact[j] ; // Cumulate for all training cases
   *gradptr++ += delta ;            // Bias activation is always 1
   }
```

If there is no hidden layer (the total number of layers is one, just the output), then we set the number of inputs feeding each output neuron to the number of inputs to the model. This does not include the bias, which is handled separately. And for the vector of values feeding each output neuron, we just grab the input case. But if there is at least one hidden layer, then the number of feeds coming into each output neuron is the number of neurons in the last hidden layer, and the feeds are these neurons' activations.

We set gradptr to point to the start of the output gradient in the grand gradient vector, grad.

For each output neuron we get its delta, which was computed in the block of code presented just before this one. Then we apply Equation 2-3 to each weight for that output neuron. The bias feed is assumed to have a constant value of 1.0, so its gradient term does not involve a prior-layer activation.

We're done with the output gradient, so we get ready to step back one layer by setting nnext, the number of neurons in the layer after the one being done, and by setting nextcoefs to the coefficient matrix for that next layer.

```
nnext = ntarg ;                    // Prepare for moving back one layer
nextcoefs = final_layer_weights ;
```

We now process all hidden layers, working backward from the last hidden layer to the first. Here is the code that will do this, and a discussion follows:

```
for (ilayer=n_all-2 ; ilayer>=0 ; ilayer--) { // For each hidden layer, working backwards
   nthis = nhid_all[ilayer] ;      // Number of neurons in this hidden layer
   gradptr = grad_ptr[ilayer] ;    // Point to gradient for this layer
   for (i=0 ; i<nthis ; i++) {     // For each neuron in this layer
      delta = 0.0 ;
```

```
    for (j=0 ; j<nnext ; j++)        // Equation 2-5 on Page 11, or sum in Equation 2-6 on Page 12
        delta += this_delta[j] * nextcoefs[j*(nthis+1)+i] ;
    delta *= hid_act[ilayer][i] * (1.0 - hid_act[ilayer][i]) ; // Derivative
    prior_delta[i] = delta ;         // Save it for the next layer back
    if (ilayer == 0) {               // First hidden layer?
        prevact = input + icase * max_neurons ; // Point to this sample
        for (j=0 ; j<n_model_inputs ; j++)
            *gradptr++ += delta * prevact[j] ;
    }
    else {      // There is at least one more hidden layer prior to this one
        prevact = hid_act[ilayer-1] ;
        for (j=0 ; j<nhid_all[ilayer-1] ; j++)
            *gradptr++ += delta * prevact[j] ;
    }
    *gradptr++ += delta ;  // Bias activation is always 1
    } // For all neurons in this hidden layer
```

For each layer, we set nthis to the number of neurons in this layer and set gradptr to the gradient for this layer. We then process each neuron in this layer. The first step is to sum the effect of this neuron on the criterion through every neuron in the next layer. For the first pass through the layer loop, when the next layer is the output layer, this is Equation 2-5 or the sum in Equation 2-6. In subsequent layers, it is the summation shown in Equation 2-8. When the summation is complete, we multiply by the derivative of the activation function (Equation 2-4) to complete Equation 2-6 or Equation 2-8. This delta must be saved in prior_delta because it will be needed for processing the next layer back. Of course, this saving is a minor waste when we are working on the first hidden layer because then there is no prior layer. But the waste is trivial.

After delta is computed, we complete Equation 2-7 (for the last hidden layer) or Equation 2-9 (for all earlier hidden layers) by multiplying by the activations feeding the current neuron. If this is the first hidden layer, these "activations" are the inputs to the model. Otherwise, they are the activations of the layer feeding the current layer. These calculations are wrapped up by handling the weight for the bias. Recall that bias activation is always 1.0 by definition.

At this point, all calculations for the current layer are complete. The deltas for the current layer were saved in prior_delta, and we now move them to this_delta where they will be used in the next pass through the loop that works backward through layers. We set nnext (the number of neurons in the next layer forward) and nextcoefs (their weights) and then loop back. This subroutine returns the criterion that will be minimized, either the MSE or the negative log likelihood.

```
for (i=0 ; i<nthis ; i++)        // These will be delta for the next layer back
  this_delta[i] = prior_delta[i] ;

  nnext = nhid_all[ilayer] ;        // Prepare for the next layer back
  nextcoefs = weights_opt[ilayer] ;
  } // For all layers, working backwards

 } // for all cases

return error ; // MSE or negative log likelihood
```

Weight Penalties

There are at least two reasons why weights in the model often tend to become overly large, and hence one usually should discourage this effect. First is the fact that inputs to a layer may be seriously collinear. For example, suppose we have two inputs and a desired output. The cases may look like this:

X1	X2	Y
1	2	3
2	4	6
3	6	9
4	8	12
5	10	15

Suppose our linear combination looks like this:

$$Y = aX_1 + bX_2 + c \qquad (2\text{-}17)$$

It is immediately apparent that $a=1$; $b=1$; $c=0$ is a solution that provides exactly the outcome we want. But a little thought reveals that there are an infinite number of solutions, among which we can find $a=-1,999,999,999,997$; $b=1,000,000,000,000$; $c=0$. Collinear variables (which may occur in the raw data, activations of hidden neurons, or both) produce enormous instability in most training algorithms. Not good.

The other reason is that large magnitude weights are often the hallmark of overfitting, the process of teaching the model to explain noise in the data in addition to legitimate patterns. Suppose the data consists of significant real patterns that are contaminated with random noise. As training begins, with weights near zero, the algorithm will probably gravitate toward weights that explain the patterns. But as training continues and the legitimate patterns become fully explained by the model, the algorithm will shift its focus to the noise components, which of course will not repeat when the model is presented with data it has not yet seen. To fit the vagaries of random noise, weights will tend to grow large. Thus, if we can prevent this increase in weights, we will likely create a model that will generalize better than a model that has larger weights that enable the model to explain the random noise in the training data.

A simple yet effective method for discouraging large magnitude weights is to add the sum of their squares to the optimization criterion. This has at least three benefits, listed here:

- It clearly accomplishes the task of discouraging large weights because we are minimizing the criterion.

- The larger a weight becomes, the more impact small changes in its weight produce. This, at least intuitively, is better than a linear penalty such as absolute value. Large weights incur an enormous penalty so that weights are allowed to become modestly large with little penalty, but truly large weights become nearly impossible.

- The sum of squares penalty is differentiable, meaning that it can be incorporated into the gradient easily.

At the risk of seeming overly pedantic, the penalty shown in Equation 2-18 has the partial derivative shown in Equation 2-19. *PenFac* is the user's specified penalty factor, typically quite small, such as 0.001 or so.

$$\textbf{\textit{Penalty}} = \textbf{\textit{PenFac}} \sum_i \textbf{\textit{w}}_i^2 \tag{2-18}$$

$$\frac{\partial \textbf{\textit{Penalty}}}{\partial \textbf{\textit{w}}_i} = \textbf{\textit{2}} \cdot \textbf{\textit{PenFac}} \cdot \textbf{\textit{w}}_i \tag{2-19}$$

To implement this weight penalty, simply add the penalty to the optimization criterion and add its derivative to each corresponding element of the gradient vector.

One generally should not include the bias terms in the weight penalty, although there may be special situations in which it is necessary. This is because a large bias may be needed to compensate for large offsets in the inputs to the network. If you know that the inputs will always be centered near zero, then bias penalties may be appropriate.

Multithreading Gradient Computation

This section is of interest to only those who want to program the relatively complex but extremely efficient threaded version of gradient computation. Most modern processors contain multiple cores that can execute code simultaneously. Training time can be greatly reduced if the expensive process of computing the gradient can be split across these multiple cores and run concurrently. The discussion here is directed at Windows applications, although programmers working with other platforms can likely adapt the basic principles appropriately.

Subroutines launched as a thread take only a single parameter, a (void *). Therefore, the programmer must embed all required parameters in a structure and pass a pointer to it. The structure used here and the threaded routine are shown next. Note that the variable names are identical to those used in the gradient routine just discussed to reduce possible confusion.

```
typedef struct {
   int istart ;
   int istop ;
   int classifier ;
   int n_all ;
   int n_all_weights ;
   int n_model_inputs ;
   int ntarg ;
   int *nhid_all ;
   int max_neurons ;
   double *input ;
   double *targets ;
   double *outputs ;
```

```
  double **weights_opt ;
  double **hid_act ;
  double *this_delta ;
  double *prior_delta ;
  double **grad_ptr ;
  double *final_layer_weights ;
  double *grad ;
  double error ;
} GRAD_THR_PARAMS ;
static unsigned int __stdcall batch_gradient_wrapper ( LPVOID dp )
{
((GRAD_THR_PARAMS *) dp)->error = batch_gradient (
                ((GRAD_THR_PARAMS *) dp)->istart ,
                ((GRAD_THR_PARAMS *) dp)->istop ,
                ((GRAD_THR_PARAMS *) dp)->input ,
                ((GRAD_THR_PARAMS *) dp)->targets ,
                ((GRAD_THR_PARAMS *) dp)->n_all ,
                ((GRAD_THR_PARAMS *) dp)->n_all_weights ,
                ((GRAD_THR_PARAMS *) dp)->n_model_inputs ,
                ((GRAD_THR_PARAMS *) dp)->outputs ,
                ((GRAD_THR_PARAMS *) dp)->ntarg ,
                ((GRAD_THR_PARAMS *) dp)->nhid_all ,
                ((GRAD_THR_PARAMS *) dp)->weights_opt ,
                ((GRAD_THR_PARAMS *) dp)->hid_act ,
                ((GRAD_THR_PARAMS *) dp)->max_neurons ,
                ((GRAD_THR_PARAMS *) dp)->this_delta ,
                ((GRAD_THR_PARAMS *) dp)->prior_delta ,
                ((GRAD_THR_PARAMS *) dp)->grad_ptr ,
                ((GRAD_THR_PARAMS *) dp)->final_layer_weights ,
                ((GRAD_THR_PARAMS *) dp)->grad ,
                ((GRAD_THR_PARAMS *) dp)->classifier ) ;
  return 0 ;
}
```

I'll now show the routine that the user calls, one small section at a time. This routine is a member of the Model class, so most of its variables are members of the class rather than being passed in its calling list. However, almost without exception, their names are identical to the names we've been using all along, so there should be no confusion. Readers wishing to more or less copy the code, should download it from the author's web site. The only reason for presenting the code in this text is to facilitate longer explanations than can be presented in comments within the code. Of necessity, niceties such as memory allocation and error checking must be omitted from the text.

The calling list and local variable declarations are shown next. Also, we divide the user's weight penalty factor by the number of weights to make it per weight and hence more intuitive.

```
double Model::gradient_thr (
  int nc ,          // Number of cases
  double *input ,   // Inputs, nc rows and max_neurons columns; n_model_inputs used
  double *target ,  // Targets, nc rows and ntarg columns
  double *grad      // Concatenated gradient vector which is computed here
  )
{
  int i, j, ilayer, ineuron, ivar, n, istart, istop, n_done, ithread ;
  int n_in_batch, n_threads, ret_val, nin_this_layer ;
  double error, *wptr, *gptr, factor ;
  double*hid_act_ptr[MAX_THREADS][MAX_LAYERS] ;
  double *grad_ptr_ptr[MAX_THREADS][MAX_LAYERS] ;
  double wpen, penalty ;
  char msg[256] ;
  GRAD_THR_PARAMS params[MAX_THREADS] ;
  HANDLE threads[MAX_THREADS] ;

  wpen = TrainParams.wpen / n_all_weights ;
```

Earlier we referred to grad_ptr as the vector of pointers to the gradient for each layer. Here is where we build this vector of pointers.

```
  gptr = grad ; // This was allocated n_all_weights * max_threads long
  for (ilayer=0 ; ilayer<n_all ; ilayer++) {
    grad_ptr[ilayer] = gptr ;
```

```
if (ilayer == 0 && n_all == 1) {          // Direct input to output? (no hidden layer)
  n = ntarg * (n_model_inputs+1) ;        // This many inputs to each neuron in layer
  gptr += n ;                             // Not needed, but it illustrates the process
  }
else if (ilayer == 0) {                   // First hidden layer?
  n = nhid_all[ilayer] * (n_model_inputs+1) ;  // This many ins to each neuron in layer
  gptr += n ;
  }
else if (ilayer < n_all-1) {              // Subsequent hidden layer?
  n = nhid_all[ilayer] * (nhid_all[ilayer-1]+1) ;  // This many ins to each neuron in layer
  gptr += n ;
  }
else
  n = ntarg * (nhid_all[ilayer-1]+1) ;    // Not needed but illustrates the process
  } // For all layers, including output
```

We now initialize the structure through which parameters will be passed to the gradient routine. These are the members that will remain unchanged throughout execution of this routine. Many of them are constants. But some of them are work vectors, and we need a separate such vector for each thread. They must be allocated sufficiently long.

Also note that hid_act and grad_ptr require special treatment because they are arrays of pointers to work areas. So, their elements need to be initialized before the pointer to the array is placed in the parameter structure. Particularly subtle, grad is allocated n_all_weights long times the number of threads, which is max_threads here. If what's going on with it in the following code is not clear, look back at the prior code block, in which grad_ptr is built from grad. This array of pointers is copied into the parameter structure, with each copy offset by the total number of weights.

```
for (i=0 ; i<max_threads ; i++) {
  params[i].input = input ;
  params[i].targets = targets ;
  params[i].n_all = n_all ;
  params[i].n_all_weights = n_all_weights ;
  params[i].n_model_inputs = n_model_inputs ;
  params[i].ntarg = ntarg ;
```

```
params[i].nhid_all = nhid_all ;
params[i].max_neurons = max_neurons ;
params[i].weights_opt = weights_opt ;
params[i].final_layer_weights = final_layer_weights ;
params[i].this_delta = this_layer + i * max_neurons ;
params[i].prior_delta = prior_layer + i * max_neurons ;
params[i].outputs = outputs + i * ntarg ;
params[i].grad = grad + i * n_all_weights ;
for (j=0 ; j<n_all ; j++) { // n_all is the number of layers, including the output layer
   hid_act_ptr[i][j] = hid_act[j] + i * max_neurons ;
   grad_ptr_ptr[i][j] = grad_ptr[j] + i * n_all_weights ;
   }
params[i].hid_act = hid_act_ptr[i] ;
params[i].grad_ptr = grad_ptr_ptr[i] ;
params[i].classifier = classifier ;
}
```

The DEEP program sets an absolute upper limit on the number of threads possible. By default, the global variable max_threads is initialized to this value, although the user has the option of reducing it. The training set will be divided into as many batches as there will be threads running, with each batch beginning with index istart and ending just before index istop. The code to start the threads is shown next. The error handling code is omitted for clarity; in the unlikely event that beginthreadex() fails, we should close all threads that have already begun and return a special error code to the caller to indicate that an unexpected and serious error has occurred.

```
n_threads = max_threads ;        // Try to use as many as possible
if (nc / n_threads < 100)        // But because threads have overhead
   n_threads = 1 ;               // Avoid using them if the batch is small

istart = 0 ;                     // Batch start = training data start
n_done = 0 ;                     // Number of training cases done in this epoch so far

for (ithread=0 ; ithread<n_threads ; ithread++) {
   n_in_batch = (nc - n_done) / (n_threads - ithread) ; // Cases left / batches left
   istop = istart + n_in_batch ;     // Stop just before this index
```

```
params[ithread].istart = istart ;   // Tell the threaded routine which
params[ithread].istop = istop ;     // training cases to process

threads[ithread] = (HANDLE) _beginthreadex (
                NULL , 0 , batch_gradient_wrapper , &params[ithread] , 0 , NULL ) ;

n_done += n_in_batch ;
istart = istop ;
} // For all threads / batches
```

We now wait for all of these threads to finish, as shown in the code on the next page. The timeout parameter, 1200000 milliseconds here, is an important consideration. If even one thread is still running when this limit is hit, the process will fail, in which case we must return a catastrophic error flag. So, we should be inspired to make this limit large. But if it is made too large, the user could be left staring at an unresponsive screen for a very long time.

```
ret_val = WaitForMultipleObjects ( n_threads , threads , TRUE , 1200000 ) ;

CloseHandle ( threads[0] ) ;
for (ithread=1 ; ithread<n_threads ; ithread++) {
  params[0].error += params[ithread].error ;
  for (i=0 ; i<n_all_weights ; i++)
    params[0].grad[i] += params[ithread].grad[i] ;
  CloseHandle ( threads[ithread] ) ;
  }

factor = 1.0 / (nc * ntarg) ;
error = factor * params[0].error ;
for (i=0 ; i<n_all_weights ; i++)
  grad[i] = factor * params[0].grad[i] ;   // Note: grad and params[0].grad are the same!
```

The previous code illustrates a standard method for pooling results from multiple threads. Each thread has its own copy of the scalar error and the vector grad. We simply add them all, starting with thread 1, to the values in thread 0. Then we divide by the number of cases and the number of targets (output neurons) to normalize results to a "per-case, per-output" scale. Of course, the thread handles must be closed.

The final step is to implement weight penalties as described on page 27. This is done in two blocks of code. The first block handles all hidden layers. Observe how grad_ptr and weights_opt supply the pointers to the start of the gradient and weights respectively for each layer. The user's weight penalty factor is applied to each gradient but applied to the error only when finished to save a bit of time.

Recall that this gradient routine actually computes the negative of the gradient, as we are minimizing. So for applying the weight penalty to the gradient terms, we flip the sign of Equation 2-19.

This code does one thing of arguable wisdom: it does not include the bias weights in the penalty. Some experts argue that because these are just constant offsets, not weights applied to inputs coming into a neuron, they should be allowed to take whatever values the training algorithm desires. This is good if the inputs to the model may themselves have large offsets from zero. But it may also be argued that if bias weights are allowed to grow without restraint, the result may be that one or more neurons remain permanently switched on or off at near saturation, rendering them useless. In any case, it's usually a moot point if the user simply makes sure that the inputs to the model are well behaved (centered reasonably near zero with modest variance and tails that are not obnoxiously heavy). I choose to not penalize bias weights because although my inputs are under control (standardized if no unsupervised layer, 0-1 if fed by an unsupervised layer), all inputs from an unsupervised layer are non-negative and there may be a lot of them. If you would prefer to penalize bias weights, just change the < loop test to <= in ivar<nin_ this_layer, and the bias weight, which is last in each neuron's weight set, will be included for both the penalty and the derivative.

```
penalty = 0.0 ;
nin_this_layer = n_model_inputs ;
for (ilayer=0 ; ilayer<n_all-1 ; ilayer++) { // Do all hidden layers
  for (ineuron=0 ; ineuron<nhid_all[ilay er] ; ineuron++) {
    wptr = weights_opt[ilayer] + ineuron*(nin_this_layer+1) ; // This neuron in this layer
    gptr = grad_ptr[ilayer] + ineuron*(nin_this_layer+1) ;
    for (ivar=0 ; ivar<nin_this_layer ; ivar++) {
      penalty += wptr[ivar] * wptr[ivar] ;
      gptr[ivar] -= 2.0 * wpen * wptr[ivar] ;
      }
    }
  nin_this_layer = nhid_all[ilayer] ;
  }
```

```
for (ineuron=0 ; ineuron<ntarg ; ineuron++) {
    wptr = final_layer_weights + ineuron * n_final_layer_weights ;
    gptr = grad_ptr[n_all-1] + ineuron * n_final_lay er_weights ;
    for (ivar=0 ; ivar<nin_this_layer ; ivar++) {
        penalty += wptr[ivar] * wptr[ivar] ;
        gptr[ivar] -= 2.0 * wpen * wptr[ivar] ;
        }
    }

return error + wpen * penalty ;
}
```

Gradient Computation with CUDA

Many excellent books have been written on the subject of general CUDA programming; this is not one of them. It would be impossible to do justice to such a broad and important topic in a few sections of a book whose primary focus is deep belief nets. However, we will attempt to satisfy these two goals:

- For the at least moderately experienced CUDA programmer, this book will present ready-to-use efficient source code, along with explanations of its motivation and operation.

- For readers with no experience in parallel programming, this book will provide an intuitive overview of neural network CUDA programming techniques, enough to satisfy curiosity and enable a decision as to whether the reader should pursue this fascinating subject.

For the sake of the latter category of readers, I should emphasize that most modern NVIDIA GPUs are CUDA-capable, and the CUDA software development kit, available for free from the NVIDIA web site, is fast and easy to install. Also, the documentation is excellent. So, do not be afraid to tackle this subject; it's easier than you may think.

Basic Architecture

The architecture of a CUDA-enabled display adapter GPU is far too complex to treat in detail here. However, so that readers new to CUDA can understand the code that will follow, we will present a basic outline of the subject.

A CUDA device runs many threads concurrently, and the programmer needs a logical way to address these threads to facilitate communication of data to and from each thread. What complicates this process is that the architecture generally looks different from the perspective of the hardware and the software. Truly expert programmers who strive for optimal performance can devote enormous effort to tuning their program so that the software addressing scheme cooperates tightly with the hardware.

Adding to the confusion is the fact that NVIDIA encourages programmers to, as far as possible, disregard much of the hardware perspective and design programs to be independent of the hardware on which they will be run. For this reason, all of the code in this book will take the approach of being hardware-independent, even though this means that performance may vary on different generations of CUDA hardware. Experts should feel free to tune the code to targets as desired.

From the software perspective, threads are grouped at two levels, a grid of *blocks* and the individual *threads* within a block. Individual blocks within the grid, as well as individual threads within a block, can be addressed in one of three ways, at the discretion of the programmer. They can be addressed linearly, from the first (0) to the last. Or they can be addressed in two dimensions, with x and y coordinates. Finally, blocks and threads can be addressed in three dimensions, with x, y, and z coordinates. The choice of addressing scheme, as well as the size of each dimension in multidimensional addressing, are and should be (if the programmer is to conform to recommendations) independent of hardware issues such as the number of multiprocessors on the device and the number of computation cores on each multiprocessor.

There is one vital aspect of hardware that must be considered when designing the program. The threads in a block are grouped into sets of 32 threads, called *warps*. (This warp size is not guaranteed by nVidia, but it has always been 32, and nVidia has implies that for the foreseeable future it will remain at 32.) Every thread in a warp is executed simultaneously. The implication is that if a warp contains fewer than 32 threads that perform useful computation, efficiency suffers. Those unused threads take up valuable resources but perform no work. This is almost the only significant hardware consideration that will play a role in program design in this text.

Figure 2-2 illustrates one possible software-defined way of addressing the threads in an application. Individual threads are shown only in the upper-left block, although they would be duplicated in all blocks.

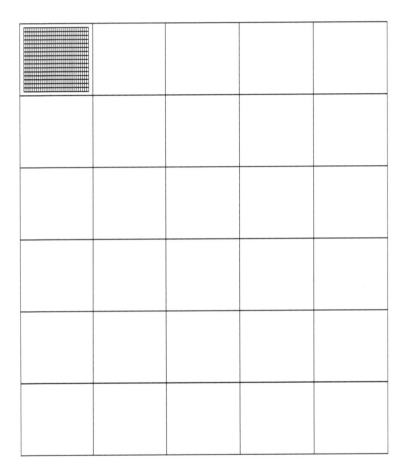

Figure 2-2. *An example CUDA grid*

In the example grid layout shown in Figure 2-2, we have a total of 30 blocks, which will be addressed in two dimensions: six rows and five columns. Each block contains 512 threads, also arranged in two dimensions of 16 rows and 32 columns. The fact that the number of columns equals the warp size is just coincidence; this need not be the case. But blocks will always be executed with an integer number of warps, whether the design specifies this or not. So, for example, suppose the programmer chose to use 513 threads. This would be bumped up to 512+32=544 threads in actual execution, with 31 of them sitting idle while nevertheless taking up resources.

A Simple Example

In the next section, we'll take up the topic of gradient computation in earnest and in a logical order. But first we'll present the simplest algorithm involved in gradient computation: transfer of the "prior" delta vector to the "current" delta vector for each case in a batch. On page 27 we saw it done for a single case with the following two trivial lines of code:

```
for (i=0 ; i<nthis ; i++)        // These will be delta for the next layer back
    this_delta[i] = prior_delta[i] ;
```

To do this in CUDA, we need two separate routines. One of them is the code that runs on the CUDA device, and the other is the host routine that will be called to invoke the device routine. Here is the device code:

```
__global__ void device_move_delta (
  int nhid      // Number of neurons in the layer just processed
)
{
  int icase, ihid ;

  ihid = blockIdx.x * blockDim.x + threadIdx.x ;

  if (ihid >= nhid)
    return ;

  icase = blockIdx.y ;

  d_this_delta[icase*nhid+ihid] = d_prior_delta[icase*nhid+ihid] ;
}
```

The keyword __global__ tells the compiler that this code will run on the CUDA device. We pass as a parameter the number of deltas that will need to be moved for each case. This is the number of neurons in the layer just processed. Each thread will move a single value, the delta for a particular neuron and case.

As is virtually always the case, the device code makes use of several built-in variables. In this example, they are as follows:

blockIdx.x: The x coordinate in the grid of the block being executed. In Figure 2-2 this would be the column (zero through four because there are five columns).

blockDim.x: The number of threads per block in this application because it uses linear thread addressing (which will become clear when the host routine is presented). If the application used multidimensional thread addressing, this would be the size of the x dimension of this block. But with linear thread addressing (and in Figure 2-2), this is the total number of threads in each block, 512.

threadIdx.x: The index of the thread being executed because linear thread addressing is being used in this application. If the thread layout were multidimensional, this would be the x coordinate of the thread being executed within the block. But with linear thread addressing this would be the linear thread ID (0 through 511 because there are 512 threads per block) of the thread in the block.

blockIdx.y: The y coordinate of the block being executed. In Figure 2-2 this would be the row (zero through five because there are six rows).

To understand the addressing used in this small example, suppose we have 64 threads per block (blockDim.x=64, not the 512 of Figure 2-2), and suppose we have 200 neurons, meaning that we will need 200 threads for each case, with each thread moving a single delta value. Then neurons 0 through 63 will be handled by block 0, neurons 64–127 by block 1, and so forth. In particular, the neuron handled by a given thread in a given block is the block number times the number of threads per block, plus the thread number. This leads to the first line of code in the routine shown earlier:

```
ihid = blockIdx.x * blockDim.x + threadIdx.x ;
```

But because there will in general be (ideally just *slightly*) more threads than neurons and the neuron index is used to address neurons within the delta vectors, we must return immediately if a thread is assigned to a neuron that does not exist.

Finally, we get the case index as the row in the grid of blocks and use the combination of the neuron index and the case index to address the move of a single delta value.

Astute readers will notice a source of inefficiency in the addressing scheme presented earlier. Unless the number of neurons happens to be an exact multiple of the number of threads per block, the last block of every row (case) will be incomplete; some threads will return immediately without doing any work but will still need to be scheduled and take up resources. We could reduce this inefficiency by addressing in the following manner:

```
index = blockIdx.x * blockDim.x + threadIdx.x ;
icase = index / nhid ;
ihid = index % nhid ;

if (icase >= ncases)
   return ;
```

This way, every block except the last is guaranteed to be full. But this would usually introduce memory alignment problems that cause even worse inefficiency. Moreover, there are at least three reasons why this problem may not be as serious as one might think.

- Unused warps within a block are not nearly as problematic as unused threads within a warp. The nearly immediate return causes completely unused warps to zip in and out of a processor in the blink of a gnat's eye, so their cost is negligible. It's partially filled warps that are the time problem because the entire warp must remain resident, even if only a few threads are executing.

- These inefficiencies are most pronounced in small tasks, when the idle threads are a high percentage of the total task. But in such cases, the time taken by the task is likely small relative to the entire application. Large tasks, whose time really matters, have only a small fraction of their threads idle.

- This is the most important reason. It's easy to think that as long as a warp is resident in a multiprocessor, it's executing. But it may stall from any of several causes. The hardware has a limited number of pipelines, such as for arithmetic, memory transfers, and so on. Also, global memory accesses may cause stalls while waiting for memory requests to be serviced. So, having the full number of threads resident and ready to execute will often overburden computational resources to the point that they stall anyway. In fact, nVidia's *Best Practices Guide* says that once you hit an occupancy rate over 50 percent, you are at the point of rapidly diminishing returns because of stalls from resource scarcity.

Now that we've seen the tiny module that runs on the CUDA device, we can examine the host routine that invokes it. This code fragment is as follows:

```
int warpsize, threads_per_block ;
dim3 block_launch ;

threads_per_block = (nhid + warpsize - 1) / warpsize * warpsize ;
if (threads_per_block > 4 * warpsize)
   threads_per_block = 4 * warpsize ;

block_launch.x = (nhid + threads_per_block - 1) / threads_per_block ;
block_launch.y = ncases ;
block_launch.z = 1 ;

device_move_delta <<< block_launch , threads_per_block >>> ( nhid ) ;
cudaDeviceSynchronize() ;
```

In this code, dim3 is a built-in vector type that is used when we want to specify a multidimensional launch parameter rather than a scalar, and warpsize was earlier set to 32 in the program. This is the hardware-fixed number of threads in a warp.

The first executed line in the previous code sets threads_per_block equal to the smallest number possible to handle all nhid neurons and be an integer multiple of the warp size.

Of course, if there are many neurons, the value just computed may exceed hardware or efficiency limits on the number of threads per block, the block size. The next two lines set an upper limit on the block size. The exact manner of doing so is not critical as long as the following basic precautions are taken:

- The final value should be an integer multiple of the warp size. Doing otherwise, while legal, guarantees unfilled warps, which is a silly waste.

- If the device code has a lot of register use, too many threads per block may exceed hardware limits. The register use of the routines in this text is small enough that there is a lot of leeway on block size. Having four warps, used here, is not even close to problematic.

- If the block size is too small, the hardware limit on the number of blocks that can run on each multiprocessor will impose an undue restraint on the number of warps eligible to execute, which will under-utilize the hardware. Again, using a block size of four warps is far from problematic on current and near-term foreseeable CUDA hardware.

The x dimension of the grid, block_launch.x, which handles the neuron index, is computed as the minimum number of blocks such that we'll have enough threads to handle every neuron. The y dimension is the number of cases, and the z dimension is one because we do not process blocks in that third dimension.

Finally, we run the device code with the following lines:

```
device_move_delta <<< block_launch , threads_per_block >>> ( nhid ) ;
cudaDeviceSynchronize() ;
```

The first line names the device routine, specifies the launch parameters, and provides the parameter list (just one variable here). The second line halts execution on the host until the device code completes. If this command were not present, the host could go on to launch another stage of computation dependent on these results before this routine was finished, an obvious disaster.

Initialization

Many things need to be done before we can begin using the CUDA device to compute the gradient whenever it is called upon to do so from the training algorithm. Most notably, these include allocating global memory on the device and allocating host memory for translating between the host's double precision and the single-precision floats used for most storage on the device. The initialization code is much too long to list here, but much of it is redundant, so we'll discuss only a couple of items. The complete CUDA code is available for free download from the author's web site.

We begin with the complete set of variable declarations. This is unfortunately long, but it is handy to have it available for reference as the various routines are discussed. Several aspects of these declarations are noteworthy.

- They are all declared to reside in the read-only (for device code) constant memory space, which has faster access than global memory.

- By my own convention, variables that begin with h_ are accessed by the host, and variables that begin with d_ are accessed by the device. If their names are otherwise identical, the variables will refer to the same quantity.

```
static float *fdata = NULL ;
static int n_hid_weights ;          // Total number of hidden weights across all layers
static int n_out_weights ;          // Total number of output weights

static float *reduc_fdata = NULL ;

__constant__ int d_ncases ;         // Number of cases in complete training set
__constant__ int d_n_trn_inputs ;   // Number of first-layer inputs (training data)
__constant__ int d_ntarg ;          // Number of targets (output neurons)

static    int *h_nhid = NULL ;      // Number of neurons in each of the hidden layers
__constant__ int *d_nhid ;

static    float *h_trn_data = NULL ; // Raw training data; ncases by n_trn_inputs
__constant__ float *d_trn_data ;

static    float *h_targets = NULL ;  // Target data; ncases by ntarg
__constant__ float *d_targets ;

static    int *h_class = NULL ;      // If classification (SoftMax), class id is here
__constant__ int *d_class ;

static    float *hidden_weights = NULL ; // Weight matricies for hidden layer
static    float **h_whid = NULL ;
__constant__ float **d_whid ;

static    float *h_wout = NULL ;
__constant__ float *d_wout ;

static    double *activations = NULL ; // Activations of this layer, which we compute
static    double **h_act = NULL ;   // Array of pointers to each layer
__constant__ double **d_act ;

static    double *h_output = NULL ; // Output activations
__constant__ double *d_output ;
```

```
static    float *h_mse_out = NULL ;
__constant__ float *d_mse_out ;

static    double *h_this_delta = NULL ; // Delta for current layer
__constant__ double *d_this_delta ;

static    double *h_prior_delta = NULL ; // Delta for next layer back
__constant__ double *d_prior_delta ;

static    int h_gradlen ;             // Length of complete gradient for a case
__constant__ int d_gradlen ;
static    float *h_gradient = NULL ;    // Gradient for all layers, including output
__constant__ float *d_gradient ;
static    float **h_grad_ptr = NULL ;  // Pointers to locations in gradient for each layer
__constant__ float **d_grad_ptr ;
```

One important memory allocation is that for the hidden-layer weights. (The output layer weights are kept separately.) Each layer's weights are in a separate array, although on the device these arrays are contiguous. An array of the pointers to each layer's weights allows fast access.

Here is the code for allocation of hidden-layer weights, and an explanation will follow:

```
  n_hid_weights = 0 ;
  n_prior = n_inputs ;

  for (i=0 ; i<n_layers-1 ; i++) {
    n_hid_weights += nhid[i] * (n_prior + 1) ; // Include bias via +1
    n_prior = nhid[i] ;
    }

  memsize = n_hid_weights * sizeof(float) ;
  cudaMalloc ( (void **) &hidden_weights , (size_t) memsize ) ;

  memsize = (n_layers-1) * sizeof(float *) ;
  cudaMalloc ( (void **) &h_whid , (size_t) memsize ) ;

  cudaMemcpyToSymbol ( d_whid , &h_whid , sizeof(void *) , 0 ,
                       cudaMemcpyHostToDevice ) ;
```

```
n_hid_weights = 0 ;
n_prior = n_inputs ;

for (i=0 ; i<n_layers-1 ; i++) {
  fptr[i] = hidden_weights + n_hid_weights ;
  n_hid_weights += nhid[i] * (n_prior + 1) ;
  n_prior = nhid[i] ;
  }

cudaMemcpy ( h_whid , &fptr[0] , (n_layers-1) * sizeof(float *) ,
              cudaMemcpyHostToDevice ) ;
```

The first step is to compute the total number of hidden-layer weights and the storage space needed for them. The weights will be stored as floats rather than doubles to conserve memory. Only quantities that propagate from layer to layer are stored as doubles as we do not want rounding errors in them to cumulate. We allocate hidden_weights to hold the complete set of all network weights.

Next we allocate the small amount of memory for the array that holds the pointers to the weights for each layer. There are n_layers total, but this includes the output layer, so we subtract one. The cudaMemcpyToSymbol call copies the address of this pointer vector, in h_whid, to the location on the device from which it will be accessed by the device routines.

We then fill in this pointer array with the locations within hidden_weights of the weights for each layer. The final step is to copy this vector of pointers to the device.

Copying the weights from the training algorithm to the device is not part of initialization; it will be done frequently during training. However, as long as we just allocated space for the weights, we'll now show the code for copying trial weights from the host to the device.

Within a layer, the weight matrix is kept as the transpose of the weight matrix on the host. On the host it is most efficient to have the inputs to each neuron vary the fastest, with each neuron having an array of input weights followed by the bias. This allows the use of fast dot product routines. But on a CUDA device it is much more efficient to have the neurons in the weight matrix vary the fastest, with each input having its own vector of weights to the neurons in the layer, and one last such vector for the bias.

The reason for this storage scheme has to do with optimal fetches of data from global memory. This will be discussed in more detail when the activation functions are presented, but the general reason is that CUDA devices (usually) fetch global memory in blocks of 128 bytes, regardless of the size of the memory request. These fetches are most efficient when most or all of the fetched data is immediately used, a process called *coalescing*. We'll see that making the weights for neurons being processed adjacent in memory facilitates good coalescing. Here is the code fragment for copying the weights, in hid_weights on the host, to the device:

```
float *fptr ;

fptr = fdata ;
n_prior = n_inputs ;

for (ilayer=0 ; ilayer<n_layers-1 ; ilayer++) {
  wptr = hid_weights[ilayer] ;
  for (ivar=0 ; ivar<=n_prior ; ivar++) {
    for (ineuron=0 ; ineuron<nhid[ilayer] ; ineuron++)
      *fptr++ = (float) wptr[ineuron*(n_prior+1)+ivar] ;
    }
  n_prior = nhid[ilayer] ;
  }

cudaMemcpy ( hidden_weights , fdata , n_hid_weights * sizeof(float) ,
             cudaMemcpyHostToDevice ) ;
```

In this code, fdata is a float scratch array already allocated. The layers are processed one at a time, with hid_weights being the pointers to the weights as they exist on the host. The weights for each layer are transferred to the scratch array with the computed neuron changing the fastest, which is the opposite of their storage on the host in which the input index changes the fastest. After all layers are done, the array of weights is copied to the device. The copying loop uses <= rather than < to include the bias weight.

Hidden Neuron Activation

The most basic operation in a multiple-layer feedforward network is computing the activation of a neuron. This was expressed in Equation 1-1. We simply find the dot product of the inputs to the neuron with the weights for this neuron. Recall that the bias is implemented as one additional weight with a constant input feed of one. Then apply the activation function, the logistic function here, to the dot product. We'll use the x coordinate of each block in the grid to define the neuron being computed, and the y coordinate will be the case number of the case. We allow the complete set of training cases to be processed in subsets so that the CUDA computation can be split into groups small enough that their processing will not freeze the video display and possibly cause Windows to complain with a forced timeout.

The code fragment for launching the neuron activation code is as follows:

```
threads_per_block = (nhid + warpsize - 1) / warpsize * warpsize ;
if (threads_per_block > 4 * warpsize)
  threads_per_block = 4 * warpsize ;

block_launch.x = (nhid + threads_per_block - 1) / threads_per_block ;
block_launch.y = istop - istart ;
block_launch.z = 1 ;
device_hidden_activation <<< block_launch , threads_per_block >>>
                          ( istart , istop , ilayer ) ;
cudaDeviceSynchronize() ;
```

In this code, nhid is the number of neurons in this hidden layer. We'll use one thread for each neuron, so exactly as in the simple example of the previous section, we compute the number of threads as the minimum warps sufficient to handle all neurons, set an upper limit, and compute the x dimension as the minimum number of blocks needed. Again, the y dimension is the number of cases being processed. We are processing the training set in groups, with istart being the starting index of the case in this group and istop being one past the last case.

The device code is shown here, and a discussion follows:

```
__global__ void device_hidden_activation (
  int istart ,    // First case in this group
  int istop ,     // One past last case
  int ilayer      // Layer to process
  )
{
  int icase, ihid, i_input, n_inputs, nhid ;
  float *f_inptr, *wptr ;
  double sum, *actptr, *d_inptr ;

  ihid = blockIdx.x * blockDim.x + threadIdx.x ;
  nhid = d_nhid[ilayer] ;
```

```
  if (ihid >= nhid)
    return ;

  icase = blockIdx.y ;

  wptr = d_whid[ilayer] ;
  actptr = d_act[ilayer] ;
  sum = 0.0 ;

  if (ilayer == 0) {
    n_inputs = d_n_trn_inputs ;
    f_inptr = d_trn_data + (icase+istart)*n_inputs ;
    for (i_input=0 ; i_input<n_inputs ; i_input++)
      sum += wptr[i_input*nhid+ihid] * f_inptr[i_input] ;
    sum += wptr[n_inputs*nhid+ihid] ; // Bias
    }
  else {
    n_inputs = d_nhid[ilayer-1] ;
    d_inptr = d_act[ilayer-1] + icase*n_inputs ;
    for (i_input=0 ; i_input<n_inputs ; i_input++)
      sum += wptr[i_input*nhid+ihid] * d_inptr[i_input] ;
    sum += wptr[n_inputs*nhid+ihid] ; // Bias
    }

  actptr[icase*nhid+ihid] = 1.0 / (1.0 + __expf(-sum)) ;
}
```

As is nearly always the case in CUDA coding, the first thing done is to get the index of whatever is being processed by the thread, which is a neuron here. Check the upper limit for this index, which here is nhid, the number of neurons in this layer. Return immediately if we are beyond the limit.

We then get the index of the case in this group, a pointer to the weights for this layer, and a pointer to the activation vector for this layer. We'll be computing the element of this vector corresponding to neuron ihid. Initialize the sum (dot product) to zero, double precision for accuracy.

If ilayer is zero, this is the first hidden layer, in which case its input vector is a training case. Otherwise, the input to this neuron is the vector of activations from the prior layer. We process each possibility separately, mainly because the training cases are floats, while the activations are doubles and hence require different pointer types. The computation is identical for both possibilities. We just sum the dot product, add in the bias term, and apply the activation function.

CUDA experts may wonder why I don't use some form of reduction to sum the dot product. Although I admit that I did not try this approach, preferring the simplicity of a direct attack, I doubt that any significant gain would be obtained. The reason is that, according to the nVidia profiler, this routine is already limited by stalls because of saturation of the arithmetic pipeline from the massive amount of double-precision math operations. This algorithm, as implemented, is already getting everything the device has to offer in the way of computational power. Besides, in real-life problems the number of cases and hidden neurons will be so large that work will already be well distributed across computation cores.

There is one vital concept to note in this code. For all of the threads within a warp and also often within a block depending on warp scheduling, the only thing that varies from thread to thread is ihid, the index of the neuron being computed. This index is primarily used to access the weight array in a loop that will be quite large if there are many inputs. Thus, it is necessary for this memory access to be as efficient as possible.

CUDA devices access cached global memory in 128-byte chunks. All 128 bytes are read in a single expensive operation, even if just one byte is needed! These chunks are always aligned on addresses that are multiples of 128 bytes. The CUDA memory allocation functions guarantee that the global memory address returned is aligned on such addresses.

Consider the first weight vector in the weight matrix. It is for the first input to the neuron, and its elements cover all neurons in the layer being processed. The first thread in the first block, ihid=0, will read the four bytes (a float) of the first element in this weight vector. *Simultaneously*, the second thread, ihid=1, will want to read the next four bytes to get the next weight, and so on. Together, the 32 threads in the warp will simultaneously want to access the first 32*4=128 bytes of the weight vector. Hey, that's the size of the cache chunk. How nice! With a single read from global memory through the cache, the device can get the weights needed by all 32 threads. The next warp will get the next 32 weights in one shot also, and so on. This is called *coalescing*, and it is a concept utterly crucial to effective CUDA programming.

Now consider what would happen if instead of storing the weights with the neuron changing the fastest, we stored them as we do in the host, with the inputs changing the fastest. Then the first thread would want the first four bytes, meaning that the full 128-byte cache line would be accessed. Unfortunately, the second thread, which is supposed to be operating concurrently with the first thread (all threads in the warp, in fact), would want a weight that is further down the line by a distance equal to four (size of float) times the number of inputs, a distance that would be much larger than 128 bytes in any but small applications. So, 124 of those 128 bytes read for the first thread are wasted. The same is true for all threads in the warp; no sharing at all would be possible. The device has limited resources for fetching global memory data, so many threads would stall as they wait for their data. This is an enormous issue that must always be taken into consideration in program design.

There is one closely related issue not handled in this code, though it's inconsequential in this application. If there are a multiple of 32 neurons, the second row of the weight matrix, that for the second input, would also line up exactly on a cache address. But if not, the first weight will lie somewhere in the middle of a cache address range, which always begins on a multiple of 128 bytes. So to access that weight, thread zero will read a chunk of 32 weights, some of which are "below" it in memory and hence worthless. Of course, the weights above it in the 128-byte chunk can be used by subsequent threads, so the loss is not huge. But when it gets to the end, some weights beyond those needed may also be discarded. There is some waste.

The usual way to solve this problem is to allocate enough extra memory at the end of each row to bring its length up to a multiple of 128 bytes. This way, each row (input for the weight matrix) is guaranteed to start at a cache address boundary, ensuring that a single global read will fetch all weights for an entire warp. The price paid is increased memory use on the device, an occasional deal killer, and somewhat increased complexity of the code.

I tried that and was surprised to see that it made no measurable difference in execution speed. The answer came when I ran the code under the nVidia profiler. It turns out that the limiting factor is arithmetic. Between the fact that multiplication and addition are needed to compute addresses within the weight vector and the cumulation of the dot product is in double precision, the arithmetic pipeline runs at nearly 100 percent capacity and is the overwhelming reason for stalls. The few additional global memory accesses, though expensive, are completely hidden by the arithmetic pipeline. They happen at their leisure while threads are waiting their turn for math.

Just so that the reader can see an example of what I am talking about when I refer to padding the weight matrix rows, I do this later when restricted Boltzmann machines are presented. But for the multiple-layer feedforward network (MLFN) code, I take the simple approach of refraining from padding in order to keep the code clear and memory requirements as small as possible. This apparently has a negligible impact on speed, and it does save valuable memory space on the device. Of course, it would be a useful exercise for the reader to modify the code for row padding. After studying the restricted Boltzmann machine code in the next chapter, the method for doing this modification should be clear.

Output Neuron Activation

The code for computing output activation is so similar to that for hidden neuron activation that it's almost not worth listing here. But to clarify a couple of key points, here it is:

```
__global__ void device_output_activation (
   int istart ,        // First case in this batch
   int n_inputs ,   // Number of inputs to the output layer, not counting bias
   int ilayer          // Hidden layer which feeds the output layer
   )
{
   int icase, iout, i_input ;
   double sum, *inptr ;

   iout = blockIdx.x * blockDim.x + threadIdx.x ;

   if (iout >= d_ntarg)
      return ;

   icase = blockIdx.y ;

   inptr = d_act[ilayer] + icase * n_inputs ;
   sum = 0.0 ;

   for (i_input=0 ; i_input<n_inputs ; i_input++)
      sum += d_wout[i_input*d_ntarg+iout] * inptr[i_input] ;
   sum += d_wout[n_inputs*d_ntarg+iout] ; // Bias

   d_output[(icase+istart)*d_ntarg+iout] = sum ;
}
```

The biggest difference is that in the last line we incorporate istart so that we can save the outputs for *all* of the training cases rather than just saving result for the current subset of the training set. This facilitates fast final processing, will be seen later.

Also, the input to the output layer will always be the activations of a hidden layer, so we get them from d_act.

SoftMax Output

The SoftMax modification of the outputs employs Equation 2-12. The code is shown here, and the fragment for invoking it follows:

```
__global__ void device_softmax (
  int istart , // First case in this batch
  int istop   // One past last case
  )
{
  int icase, iout ;
  double *outptr, sum ;

  icase = blockIdx.x * blockDim.x + threadIdx.x ;
  if (icase >= istop - istart)
    return ;

  outptr = d_output + (icase + istart) * d_ntarg ; // Output vector for this case
  sum = 0.0 ;
  for (iout=0 ; iout<d_ntarg ; iout++) {
    if (outptr[iout] < 300.0)
      outptr[iout] = __expf ( outptr[iout] ) ;
    else
      outptr[iout] = __expf ( 300.0 ) ;
    sum += outptr[iout] ;
    }

  for (iout=0 ; iout<d_ntarg ; iout++)
    outptr[iout] /= sum ;
}

-----------------------------------------------------------------------------------

  n = istop - istart ; // Number of elements in training subset
  threads_per_block = (n + warpsize - 1) / warpsize * warpsize ;
```

```
if (threads_per_block > 4 * warpsize)
  threads_per_block = 4 * warpsize ;

blocks_per_grid = (n + threads_per_block - 1) / threads_per_block ;
device_softmax <<< blocks_per_grid , threads_per_block >>> ( istart , istop ) ;
cudaDeviceSynchronize() ;
```

This code processes the istart-to-istop subset of the training set. Each thread handles a single case, and the thread index is the offset within this subset of the case to be processed.

Observe that outptr, the pointer to the output vector for this case, incorporates istart, the index in the training set of the start of the subset. This is because although the training set is processed in subsets, the predicted outputs for all training cases are saved as they are computed.

Checking the value of each output before it is exponentiated is critical. In pathological situations, computed outputs can occasionally be large enough such that their exponentiated value overflows floating-point representation. If this were to happen, the result would be a non-number, and further operations on such values would propagate the non-number status. This leads to much ugliness.

Invocation of this device code is slightly different in this routine than in prior routines. Previously, the block launch parameter was multidimensional, so it was declared as dim3 type, and we specified x, y, and z values. But in this application, thread assignment is by case only, not case and neuron. So, the block identifier needs to have only one dimension. Thus, we declare it as an integer and specify that integer in the launch, just as we have been doing with the thread parameter.

Output Delta

The routine for computing the output delta (Equation 2-2) appears next, and the fragment for its invocation follows:

```
__global__ void device_output_delta (
  int istart ,    // First case in this batch
  int istop ,    // One past last case
  int ntarg     // Number of targets (outputs)
  )
{
  int icase, iout ;
```

```
  iout = blockIdx.x * blockDim.x + threadIdx.x ;
  if (iout >= d_ntarg)
    return ;
  icase = blockIdx.y ;
  d_this_delta[icase*ntarg+iout] = 2.0 * (d_targets[(icase+istart)*ntarg+iout] -
                                          d_output[(ic ase+istart)*ntarg+iout]) ;
}
-----------------------------------
  threads_per_block = (ntarg + warpsize - 1) / warpsize * warpsize ;
  if (threads_per_block > 4 * warpsize)
    threads_per_block = 4 * warpsize ;
  block_launch.x = (ntarg + threads_per_block - 1) / threads_per_block ;
  block_launch.y = istop - istart ;
  block_launch.z = 1 ;
  device_output_delta <<< block_launch, threads_per_block >>> ( istart , istop , ntarg );
  cudaDeviceSynchronize() ;
```

First, note that the indices for the targets and outputs incorporate istart because they are all stored, while deltas are saved for only the subset.

There is one gotcha to be aware of. Usually there are few outputs, perhaps just one, which results in numerous idle threads in each warp. Hence, one might be tempted to reverse identities and use the thread index for the case and blockIdx.y for the output. But delta, the targets, and the outputs are all stored with the output changing fastest. So if there were numerous outputs, reversing identities would cause massive memory stalls, as discussed on page 50.

Output Gradient

The code for computing the output gradient is slightly more complex than we've seen before because the block layout is fully three-dimensional. We'll implement Equation 2-3. It might be helpful to turn back to page 31-32 to review how grad_ptr is constructed. Each element of this array is a pointer to the concatenated gradient vector for a layer. The device code is as follows:

```
__global__ void device_output_gradient (
  int nc ,       // Number of cases in batch
  int ilayer     // Hidden layer which feeds the output layer
  )
```

```
{
  int icase, iout, ihid, nhid ;
  float *gptr ;
  double input ;

  ihid = blockIdx.x * blockDim.x + threadIdx.x ;
  nhid = d_nhid[ilayer] ;      // Neurons in last hidden layer
  icase = blockIdx.y ;

  if (ihid > nhid)
    return ;
  else if (ihid < nhid)
    input = d_act[ilayer][icase*nhid+ihid] ;
  else
    input = 1.0 ; // Bias

  iout = blockIdx.z ;

  gptr = d_grad_ptr[ilayer+1] + icase * d_gradlen ; // Gradient of output layer
  gptr[iout*(nhid+1)+ihid] = d_this_delta[icase*d_ntarg+iout] * input ;
}
```

This code is highly instructive in how to efficiently assign thread identities. Please review the discussion of memory coalescing on page 50 if needed. We could let the thread index be the case, the output neuron, or the neuron in the last hidden layer that feeds the output layer. Look at the last line of the routine. If the thread index were the case, accesses to d_this_delta would be done in potentially large strides throughout each warp as each increment of icase for the threads running parallel in the warp increments the memory address by the number of outputs, d_ntarg. No good. And if the thread index were the output, writes through gptr in the last line would stride by nhid+1. Also bad. But ihid is used for memory access in only two places: for getting the prior layer's activity d_act and for writing the gradient in the last line. Bingo. In both cases, the memory locations are adjacent, so it's a no-brainer; both of these memory accesses are coalesced or close to it, and all other memory accesses in a warp are duplicates. Then it's a toss-up for which of the other two to assign to the y and z block indices.

The only item of note in this code is the three-way decision: if the hidden neuron index exceeds the number of neurons, we are out of range, so return immediately. In the most common case we are getting the activation of a neuron. But if this thread handles the bias term, set the "input" to the constant 1.0.

The launch code shown here is straightforward. Just note that we have nhid+1 threads because thread nhid handles the weight for the bias.

```
threads_per_block = (nhid + 1 + warpsize - 1) / warpsize * warpsize ;
if (threads_per_block > 4 * warpsize)
   threads_per_block = 4 * warpsize ;

block_launch.x = (nhid + 1 + threads_per_block - 1) / threads_per_block ;
block_launch.y = nc ;
block_launch.z = ntarg ;

device_output_gradient <<< block_launch , threads_per_block >>> ( nc , ilayer ) ;
cudaDeviceSynchronize() ;
```

Gradient of the First Hidden Layer

We now come to the most complex algorithm shown so far. Let's jump right in with the code and walk through it after. Please be sure that you fully understand the code for the output gradient, shown in the prior section. This code is an extension of that algorithm.

```
__global__ void device_first_hidden_gradient (
  int istart ,       // First case in this batch
  int istop ,        // One past last case
  int only_hidden // Is this the only hidden layer?
  )
{
  int j, icase, iin, ihid, nhid, ninp1, n_next ;
  float *gptr, *next_weights, input ;
  double *delta_ptr, this_act, delta ;

  iin = blockIdx.x * blockDim.x + threadIdx.x ;
  icase = blockIdx.y ;

  if (iin > d_n_trn_inputs)
    return ;
```

```
  else if (iin < d_n_trn_inputs)
    input = d_trn_data[(icase+istart)*d_n_trn_inputs+iin] ; // Feed coming into this layer
  else
    input = 1.0f ;                        // Bias

  ihid = blockIdx.z ;
  nhid = d_nhid[0] ;                      // Neurons in this hidden layer
  ninp1 = d_n_trn_inputs + 1 ;     // We mustn't forget the bias

  if (only_hidden) {                      // Is the next layer the output or another hidden?
    n_next = d_ntarg ;
    next_weights = d_wout + ihid * n_next ;
    }

  else {
    n_next = d_nhid[1] ;
    next_weights = d_whid[1] + ihid * n_next;
    }

  delta_ptr = d_this_delta + icase * n_next ; // Delta for this case
  delta = 0.0 ;                               // Sum for Equation 2-8
  for (j=0 ; j<n_next ; j++)
    delta += delta_ptr[j] * next_weights[j] ;

  this_act = d_act[0][icase*nhid+ihid] ;     // Complete Equation 2-8
  delta *= this_act * (1.0 - this_act) ;     // Equation 2-4

  gptr = d_grad_ptr[0] + icase * d_gradlen ; // Gradient of first hidden layer
  gptr[ihid*ninp1+iin] = delta * input ;     // Equation 2-9
}
```

The first third of this code is closely related to the output gradient code. There, the thread index was the hidden neuron feeding the output. Here, by analogy, the thread index is the input feeding this first layer. The case in this subset is the *y* dimension of the block. Still by analogy, we make a three-way choice. If the thread index exceeds the number of inputs, we are out of range and must immediately return. If it is less than the number of inputs, it is an input. And if it equals the number of inputs, then this thread handles the bias, so we set the "input" to 1.0, exactly as before. Still by analogy, the *z* dimension of the block is the hidden neuron in this layer, just as before this dimension was the output neuron.

A difference we now encounter is that this first hidden layer may be the only hidden layer, in which case it feeds the output. Or there may be one or more subsequent hidden layers, in which case this layer feeds that next hidden layer. We find the number of neurons in the layer being fed by this layer, either the output or the next hidden layer. We also get a pointer to the weights going to that next layer.

This "next layer" information is used to compute the summation in Equation 2-8. This equation is completed by multiplying by the derivative of the activation function, given by Equation 2-4.

The final step is to get the pointer to the gradient for this first hidden layer (index 0 in grad_ptr), offset it for this case in the subset being processed, and store the result of Equation 2-9.

The concept of memory coalescing is important enough to again analyze the situation here, exactly as was done for the output gradient. The thread index is iin, the input index. Observe that this index is used for memory access twice: once for getting this input from the training data and once for storing the element of the gradient computed by the thread. In both cases, increments of iin refer to successive memory locations. Thus, memory accesses will be coalesced, except at the ends if these arrays do not start at multiple of the 128-byte cache line. But as pointed out before, the effect of such a slight mismatch is negligible when the limiting factor is the arithmetic pipeline.

We end this section with the code fragment used to launch this routine. It is nearly identical to code shown before, so no special explanation is needed.

```
threads_per_block = (nin + 1 + warpsize - 1) / warpsize * warpsize ;
if (threads_per_block > 4 * warpsize)
  threads_per_block = 4 * warpsize ;

block_launch.x = (nin + 1 + threads_per_block - 1) / threads_per_block ; // Include bias
block_launch.y = istop - istart ;
block_launch.z = nhid ;

device_first_hidden_gradient <<< block_launch , threads_per_block >>>
                            ( istart , istop , only_hidden ) ;
cudaDeviceSynchronize() ;
```

Gradient of Subsequent Hidden Layers

This algorithm computes the gradient for any hidden layer other than the first. It is similar to the algorithm for the first hidden layer. However, there are enough differences to make coding it separately worthwhile.

```
__global__ void device_subsequent_hidden_gradient (
  int nc ,          // Number of cases in subset
  int ilayer ,      // Hidden layer being processed
  int last_hidden  // Is this the last hidden layer?
  )
{
  int j, icase, iin, ihid, nhid, nin, ninp1, n_nex t ;
  float *gptr, *next_weights ;
  double *delta_ptr, *prior_delta_ptr, this_act, delta, input ;

  iin = blockIdx.x * blockDim.x + threadIdx.x ;
  icase = blockIdx.y ;
  nin = d_nhid[ilayer-1] ;    // Number of inputs to each neuron in this layer

  if (iin > nin)
    return ;
  else if (iin < nin)
    input = d_act[ilayer-1][icase*nin+iin] ;
  else
    input = 1.0 ;          // Bias

  ihid = blockIdx.z ;
  nhid = d_nhid[ilayer] ;    // Neurons in this hidden layer
  ninp1 = nin + 1 ;          // We mustn't forget the bias, so nin+1

  if (last_hidden) {
    n_next = d_ntarg ;
    next_weights = d_wout + ihid * n_next ;
    }
  else {
    n_next = d_nhid[ilayer+1] ;
    next_weights = d_whid[ilayer+1] + ihid * n_next ;
    }
```

```
// Pointer to delta coming from the next layer, which was just done
   delta_ptr = d_this_delta + icase * n_next ;
// Save delta for the next layer done, one layer back
   prior_delta_ptr = d_prior_delta + icase * nhid ;

   delta = 0.0 ;                            // Sum for Equation 2-8
   for (j=0 ; j<n_next ; j++)
      delta += delta_ptr[j] * next_weights[j] ;
   this_act = d_act[ilayer][icase*nhid+ihid] ;    // Complete Equation 2-8
   delta *= this_act * (1.0 - this_act) ;         // Equation 2-4
   prior_delta_ptr[ihid] = delta ;                // Save it for the next layer back

   gptr = d_grad_ptr[ilayer] + icase * d_gradlen ;  // Gradient of this hidden layer
   gptr[ihid*ninp1+iin] = delta * input ;
}
```

We'll zip through this quickly because of the many similarities with the first hidden-layer gradient. As before, the thread index is iin, the index of the input to this layer, which was a training case for the first hidden layer but is a neuron for this subsequent hidden layer. The y index of the block is the case being processed, and the z index is the neuron under consideration. We compare iin to the number of inputs and either return (or get) a prior-layer activation or use the constant 1.0 for the bias term.

Also, as we did for the first hidden layer, we inquire as to the next layer; is it the output layer or yet another hidden layer? We find the number of neurons there and get a pointer to the weights connecting to it. Then we compute the sum for Equation 2-8 and multiply by the derivative of the activation function. But unlike for the first hidden layer, we must now save the delta vector for this layer because we will need it when we process the next layer back.

It's worth noting that once again by using the input index iin as the thread index we have good memory coalescing. I'll leave it to the reader to confirm that both memory references with iin are adjacent.

There is one more little bit of device code associated with computing the gradient for hidden layers other than the first. When we computed delta in the previous code, we saved it in d_prior_data. Every time we move back one layer we need to transfer those deltas to d_this_delta for use in summing Equation 2-8. Here is the routine for doing so, followed by the code fragment for launching these two routines:

```
__global__ void device_move_delta (
  int nhid // Number of neurons in the layer just processed
  )
{
  int icase, ihid ;

  ihid = blockIdx.x * blockDim.x + threadIdx.x ;
  if (ihid >= nhid)
    return ;

  icase = blockIdx.y ;
  d_this_delta[icase*nhid+ihid] = d_prior_delta[icase*nhid+ihid] ;
}
  threads_per_block = (nhid_prior + 1 + warpsize - 1) / warpsize * warpsize ;
  if (threads_per_block > 4 * warpsize)
    threads_per_block = 4 * warpsize ;
  block_launch.x = (nhid_prior + 1 + threads_per_block - 1) / threads_per_block ;
  block_launch.y = nc ;
  block_launch.z = nhid_this ;

  device_subsequent_hidden_gradient <<< block_launch , threads_per_block >>>
                                ( nc , ilayer , last_hidden ) ;
  cudaDeviceSynchronize() ;

  threads_per_block = (nhid_this + warpsize - 1) / warpsize * warpsize ;
  if (threads_per_block > 4 * warpsize)
    threads_per_block = 4 * warpsize ;
  block_launch.x = (nhid_this + threads_per_block - 1) / threads_per_block ;
  block_launch.y = nc ;
  block_launch.z = 1 ;

  device_move_delta <<< block_launch , threads_per_block >>> ( nhid_this ) ;
  cudaDeviceSynchronize() ;
```

Fetching the Gradient

We've seen how to compute the gradient for the output layer, the first hidden layer, and other hidden layers. Now we explore how to retrieve the gradient from the CUDA device so that the training routine running on the host computer can use it.

Recall that for each case the gradient is strung out as one long vector, with one gradient vector for each case. We perform the fetching operation in two steps. First we use the device to sum the contribution of each case into the vector for the first case, and then we copy this sum to the host.

There is no good reason to employ a sophisticated algorithm such as reduction for the summing because the act of summing and fetching the gradient is a tiny fraction of the total gradient compute time. Even a huge improvement in its speed would contribute little to the overall speed of the application. Moreover, in all cases of practical interest, the gradient vector is huge. Consider, for example, the MNIST dataset with 784 inputs. If the first hidden layer has 500 neurons, we have 500*(784+1)=392500 elements in that layer's gradient alone. This allows ample opportunity for parallel operation. Here is the gradient-fetching code:

```
__global__ void device_fetch_gradient (
  int nc        // Number of cases in batch
  )
{
  int index, icase ;
  float *gptr ;
  double sum ;

  index = blockIdx.x * blockDim.x + threadIdx.x ;
  if (index >= d_gradlen)
    return ;

  sum = 0.0 ;
  gptr = d_gradient + index ;
  for (icase=0 ; icase<nc ; icase++) // For all cases in this batch
    sum += gptr[icase*d_gradlen] ;
  *gptr = sum ;
}
```

Notice that here, as in all other routines, memory accesses are well coalesced, and perfectly so when d_gradlen happens to be a multiple of 128 bytes. This is because all memory access is through gptr, and this pointer is incremented by the thread index, ensuring that reads and writes by adjacent threads in a warp are reading and writing adjacent memory.

Also notice that the gradient is single precision to save space and computation time, but the summation is done in double precision to preserve significant digits when the magnitudes of summands differ by large amounts. The sum is rounded back to single precision when complete.

Here is the code for launching this routine and fetching the gradient:

```
int cuda_fetch_gradient (
  int nc ,          // Number of cases in batch
  double *grad     // Gradient sum output here
  )
{
  int i, warpsize, blocks_per_grid, threads_per_block ;
  char msg[256] ;

  warpsize = deviceProp.warpSize ;     // Threads per warp, likely 32 well into the future

  threads_per_block = (h_gradlen + warpsize - 1) / warpsize * warpsize ;
  if (threads_per_block > 4 * warpsize)
    threads_per_block = 4 * warpsize ;

  blocks_per_grid = (h_gradlen + threads_per_block - 1) / threads_per_block ;

  device_fetch_gradient <<< blocks_per_grid , threads_per_block >>> ( nc ) ;
  cudaDeviceSynchronize() ;

  cudaMemcpy ( fdata , h_gradient , h_gradlen * sizeof(float) ,
                cudaMemcpyDeviceToHost ) ;
  for (i=0 ; i<h_gradlen ; i++)
    grad[i] += fdata[i] ;

  return 0 ;
}
```

The launch parameters are computed in the usual way, ensuring the minimum number of blocks needed to handle all gradient elements. Then we launch the device routine, wait for it to finish, and copy the summed gradient to the host. The variable fdata is a static float array that was already allocated for multiple uses.

The final step is to sum the gradient for this subset of the training set into the host's gradient vector.

One subtle point about this code should be of interest to many readers. We call cudaDeviceSynchronize() in order to halt host execution until the device is finished with the task. In reality, this is not necessary because cudaMemcpy() blocks execution until the device is finished. Still, I make it a habit to insert the redundant cudaDeviceSynchronize() to make profiling more informative. If that call were not present, the runtime API call summary would lump device execution time in with memory copy time because we sit in the copying routine while the device runs. By synchronizing before copying, the individual times (execution and copy) are displayed correctly.

Mean Squared Error by Reduction

This section presents a fascinating and powerful algorithm for rapid parallel processing of long arrays. We need such an algorithm for computing the mean squared error of the model's predictions. Even though the training set was processed in subsets and the MSE could have been computed for each subset and summed, it turns out that it's faster to save them all and then compute the MSE at once with a very efficient algorithm. Moreover, it is always good to preserve the complete set of all predictions for post-training analysis.

The mean squared error is defined as the mean across all training cases and outputs of the squared difference between the predicted values and the actual values of the target. To make it possible to illustrate the process of computing the sum of these terms, we assume a small number of terms, just 59. Figure 2-3 shows the indices of these terms, laid out in an arrangement that will help to make the algorithm clear.

0	1	2	3	4	5	6	7
8	9	10	11	12	13	14	15
16	17	18	19	20	21	22	23
----	----	----	----	----	----	----	----
24	25	26	27	28	29	30	31
32	33	34	35	36	37	38	39
40	41	42	43	44	45	46	47
----	----	----	----	----	----	----	----
48	49	50	51	52	53	54	55
56	57	58					

Figure 2-3. *Fifty-nine cases to be processed by reduction*

In this example, we'll assume that the algorithm will employ three blocks (single rows) and eight threads (columns) per block. Naturally, in a real application there will be many blocks and a multiple of 32 threads per block. We will also soon see that the number of threads per block must be a power of two.

The algorithm has three major steps. In the first step, partial sums are cumulated in the upper set of *number of blocks* times *threads per block* threads, which is 3*8=24 here. The sums are taken in steps of this product. So, the entry in location 0 would become the sum of the values in locations 0, 24, and 48. We may run out of data, which is fine. The sums do not all need to have the same number of terms. For example, the entry in location 11 would become the sum of the values in locations 11 and 35 because there is no data in location 59. When this first step is complete, all of the data that we need will be in the first 24 locations.

The second step sums the columns of each row, leaving the partial sums in the first column. This is done in substeps, cutting the number of partial sums in half each time by summing two values. In the first substep, the value in location 0 will have the value in location 4 added to it. Similarly, we have (in ideally clear notation) the following:

1 += 5

2 += 6

3 += 7

8 += 12

and so forth through

19 += 23

This operation is repeated in the next substep.

> 0 += 2
>
> 1 += 3
>
> 8 += 10
>
> 9 += 11
>
> 16 += 18
>
> 17 += 19

One more repetition completes the second step.

> 0 += 1
>
> 8 += 9
>
> 16 += 17

At this point we have just three partial sums, one for each of the three blocks. Some implementations do the final sum on the device, but I chose to do it on the host, which is just as fast and is easier.

The device code for implementing this algorithm is as follows:

```
#define REDUC_THREADS 256
#define REDUC_BLOCKS 64

__global__ void device_mse ()
{
   __shared__ double partial_mse[REDUC_THREADS] ;
   int i, index ;
   unsigned int n ;
   double diff, sum_mse ;

   index = threadIdx.x ;
   n = d_ncases * d_ntarg ; // Number of components; cases times targets (outputs)

   sum_mse = 0.0 ;
   for (i=blockIdx.x*blockDim.x+index ; i<n ; i+=blockDim.x*gridDim.x) {
      diff = d_output[i] - d_targets[i] ;
      sum_mse += diff * diff ;
      }
```

```
  partial_mse[index] = sum_mse ;
  __syncthreads() ;

  for (i=blockDim.x>>1 ; i ; i>>=1) {
    if (index < i)
      partial_mse[index] += partial_mse[index+i] ;
    __syncthreads() ;
    }

  if (index == 0)
    d_mse_out[blockIdx.x] = partial_mse[0] ;
}
```

The *shared* declaration at the top of this code allocates extremely fast memory that is shared among all threads in a block. A separate allocation exists for each block, and they are invisible to one another.

The thread index threadIdx.x will range from 0 through one less than the number of threads per block, REDUC_THREADS.

The total number of elements (squared differences between the output and the true value) is n. The cumulation is done in double precision to avoid loss of significant digits when the components greatly differ in magnitude.

The first loop executes the first step described previously. Each thread steps through the *number of blocks* times *threads per block* elements. The former is gridDim.x, and the latter is blockDim.x. When this loop is complete, the sum is placed in the shared memory area corresponding to this thread. Recall that each block has its own private copy of shared memory, so there is no interference across blocks.

We then execute the synchthreads() function. When a thread calls this function, the thread halts its execution at that point until all threads in the block reach that same point. The execution scheduler in the device can (and does) run warps in undefined order within a resident block, and in the next step we will access the shared memory computed in this first step. To ensure that the required information is available and correct, we must wait until all threads finish their computation of the first step.

The second step does the repeated "cut partial sums in half" algorithm previously described. In the first pass through the loop, a column in the right half of the block (when visualized as in Figure 2-3) is added to a column in the left half. Again, a call to synchthreads() is needed to ensure that data needed for the next pass through the loop will be available. This merging of pairs of partial sums continues until it is down to one column.

After that step is complete, the sum of each block (row in the figure) will be in the first column, which is partial_mse[0] of that block. We store it in an array dedicated to the block sums. This array was allocated during initialization. The code shown previously uses thread 0 to do this store, but actually it could be any thread because the shared memory partial_mse is the same for all threads in the block. Of course, we have to pick one and only one to avoid an attempt at multiple threads in a block trying to store to the same location!

The host code for performing the MSE computation is shown here:

```
int cuda_mse (
  int n ,          // Number of values; ncases * ntarg
  double *mse      // Computed mse criterion
  )
{
  int i, blocks_per_grid ;
  double sum ;
  char msg[256] ;

  blocks_per_grid = (n + REDUC_THREADS - 1) / REDUC_THREADS ;
  if (blocks_per_grid > REDUC_BLOCKS)
    blocks_per_grid = REDUC_BLOCKS ;

  device_mse <<< blocks_per_grid , REDUC_THREADS >>> () ;
  cudaDeviceSynchronize() ;

  cudaMemcpy ( reduc_fdata , h_mse_out , blocks_per_grid * sizeof(float) ,
               cudaMemcpyDeviceToHost ) ;

  sum = 0.0 ;
  for (i=0 ; i<blocks_per_grid ; i++)
    sum += reduc_fdata[i] ;
  *mse = sum / n ;

  return 0 ;
}
```

As usual, we use the minimum number of blocks necessary to sum all n squared errors. Many implementations of this algorithm would invoke reduction a second time, with the block partial sums as the input and using a single block. The result would be a single value, the complete sum. However, I found it just as easy and a bit more straightforward to copy the block partial sums to the host and sum there. You pay your money and you take your choice.

As a final note, there exist some considerably more complex reduction algorithms that are marginally faster than the version used here. Because I value simplicity, I chose this straightforward version. Interested readers can find examples of other approaches in the CUDA SDK.

Log Likelihood by Reduction

Classification tasks that use a SoftMax output layer require that instead of minimizing mean squared error we minimize the negative log likelihood as per Equation 2-15. The device code for doing this is shown next. It is almost identical to the MSE code, so little explanation is needed. We use the same pre-allocated array, d_mse_out, as we used for MSE. Also note that we add 1.e−30 to the output activation just to prevent the unlikely but disastrous situation of trying to take the log of zero.

```
__global__ void device_ll ()
{
   __shared__ double partial_ll[REDUC_THREADS] ;
   int i, n, ntarg, index ;
   double sum_ll ;

   index = threadIdx.x ;
   n = d_ncases ;
   ntarg = d_ntarg ;

   sum_ll = 0.0 ;
   for (i=blockIdx.x*blockDim.x+index ; i<n ; i+=blockDim.x*gridDim.x)
      sum_ll -= log ( d_output[i*ntarg+d_class[i]] + 1.e-30 ) ;

   partial_ll[index] = sum_ll ;
   __syncthreads() ;

   for (i=blockDim.x>>1 ; i ; i>>=1) {
      if (index < i)
         partial_ll[index] += partial_ll[index+i] ;
      __syncthreads() ;
   }
```

```
  if (index == 0)
    d_mse_out[blockIdx.x] = partial_ll[0] ;
}
```

Putting It All Together

We've seen all of the individual components of CUDA gradient computation. This section will show an outline of the order in which they would be called to accomplish this task. Some components, such as error handling, are omitted because those are specific to the implementation and would just clutter this presentation. However, the code available as a free download from the author's web site does demonstrate the details of a reasonable approach.

All of the CUDA routines given in prior sections were designed so that the training set could be processed in subsets, which are sometimes referred to as *batches* in the following code. This must not be confused with the more common use of the term *batch*. In that common use, the training algorithm computes the gradient and optimization criterion in batches (subsets of the training set) and updates the weight matrix for each batch. This is *not* done here because it has been my experience that when a high-quality deterministic training algorithm is used, fastest convergence is obtained when the entire training set is processed for each gradient/criterion computation and weight update. Of course, it should be easy for programmers to modify the code that follows to employ traditional batch processing.

So, why are we breaking up the training set into separately processed subsets? The reason is that many operating systems, including Windows, allow the display adapter to be tied up with computation for only short intervals of time, generally two seconds. If computation is not complete by the time the clock runs out, the operating system unceremoniously dumps the program or inflicts other less dramatic but equally disastrous punishment. This notorious *WDDM timeout* is the bane of CUDA programmers. The most common solution is to allow computations to be broken up into chunks that are processed one at a time. This is our approach.

The essential code for calling the CUDA gradient routines appears here, and a step-by-step explanation follows:

```
double Model::gradient_cuda (
  int nc ,          // Number of cases
  double *input ,  // Training set matrix, nc by Model::n_model_inputs
  double *target , // Target matrix, nc by Model::ntarg
  double *grad     // Complete gradient
  )
{
  int i, k, n, ilayer, ineuron, ivar, ret_val, ibatch, n_in_batch, n_batches ;
  int istart, istop, n_done, max_batch ;
  int n_prior, gradlen, nin_this_layer ;
  double mse, wpen, *wptr, *gptr ;

// Set up pointers to gradient for each layer
  gptr = grad ; // Training routine allocated this

  for (ilayer=0 ; ilayer<n_all ; ilayer++) {
    grad_ptr[ilayer] = gptr ;

    if (ilayer == 0 && n_all == 1) {      // Direct input to output?
      n = ntarg * (n_model_inputs+1) ; // This many inputs to each neuron in this layer
      gptr += n ;                          // Not needed, but it illustrates the process
      }

    else if (ilayer == 0) {               // First hidden layer?
      n = nhid_all[ilayer] * (n_model_inputs+1) ; // This many inputs to each neuron
      gptr += n ;
      }

    else if (ilayer < n_all-1) {          // Subsequent hidden layer?
      n = nhid_all[ilayer] * (nhid_all[ilayer-1]+1) ; // This many inputs to each neuron
      gptr += n ;
      }

    else
      n = ntarg * (nhid_all[ilayer-1]+1) ; // Not needed but it illustrates the process

    } // For all layers, including output
```

```
/*
   In order to prevent integer overflow in allocating memory for the gradient
   we compute the minimum number of batches needed to get each batch small enough.
*/

   gradlen = 0 ;
   n_prior = n_model_inputs ;
   for (i=0 ; i<n_all-1 ; i++) { // Hidden layers
     gradlen += nhid_all[i] * (n_prior + 1) ;
     n_prior = nhid_all[i] ;
     }
   gradlen += ntarg * (n_prior + 1) ;   // Output layer

   max_batch = MAXPOSNUM / (gradlen * sizeof(float)) ; // Memory allocation size
   if (max_batch > 65535)                 // Grid dimension
     max_batch = 65535 ;
   n_batches = nc / max_batch + 1 ;

/*
   Initialize CUDA device if not yet done for this session
*/

   if (! mlfn_cuda_initialized) {

     n_done = 0 ;       // Must find max batch size for cuda init
     for (ibatch=0 ; ibatch<n_batches ; ibatch++) {
       n_in_batch = (nc - n_done) / (n_batches - ibatch) ;   // Cases left / batches left
       if (ibatch == 0 || n_in_batch > max_batch)
         max_batch = n_in_batch ;
       n_done += n_in_batch ;
       }

     mlfn_cuda_init ( ... max_batch ... ) ;
     mlfn_cuda_initialized = 1 ;
     }
```

```
  if (cuda_weights_changed) {
    cuda_weights_to_device ( ... ) ;
    cuda_weights_changed = 0 ;
    }
/*
  Gradient computation starts here
*/

  for (i=0 ; i<n_all_weights ; i++)
    grad[i] = 0.0 ;

  istart = 0 ;        // Batch start = training data start
  n_done = 0 ;     // Number of training cases done in this epoch so far

  for (ibatch=0 ; ibatch<n_batches ; ibatch++) {
    n_in_batch = (nc - n_done) / (n_batches - ibatch) ;  // Cases left / batches left
    istop = istart + n_in_batch ;                         // Stop just before this index

/*
  Forward pass
*/

    for (ilayer=0 ; ilayer<n_all-1 ; ilayer++)
      cuda_hidden_activation ( istart , istop , ... , ilayer ) ;
    cuda_output_activation ( istart , istop , ... ) ;

    if (classifier)
      cuda_softmax ( istart , istop ) ;

/*
  Backward pass
*/

    cuda_output_delta ( istart , istop , ... ) ;
    cuda_output_gradient ( n_in_batch , ... ) ;
    for (ilayer=n_all-2 ; ilayer>0 ; ilayer--)
      cuda_subsequent_hidden_gradient ( n_in_batch , ilayer , ... ) ;
    cuda_first_hidden_gradient ( istart , istop , ... ) ;
    cuda_fetch_gradient ( n_in_batch , grad ) ;
```

```
    n_done += n_in_batch ;
    istart = istop ;
    } // For all batches

  for (i=0 ; i<n_all_weights ; i++)
    grad[i] /= nc * ntarg ;

  if (classifier) {
    cuda_ll ( nc , &mse ) ;
    mse /= ntarg ; // cuda_ll() divided by n but not ntarg
    }
  else
    cuda_mse ( nc * ntarg , &mse ) ;

/*
  Deal with weight penalty
  First block of code does hidden layers, second does output layer
*/

  wpen = TrainParams.wpen / n_all_weights ;
  penalty = 0.0 ;
  nin_this_layer = n_model_inputs ;

  for (ilayer=0 ; ilayer<n_all-1 ; ilayer++) { // Do all hidden layers
    for (ineuron=0 ; ineuron<nhid_all[ilay er] ; ineuron++) {
      wptr = weights_opt[ilayer] + ineuron*(nin_this_layer+1) ; // Weights for this neuron
      gptr = grad_ptr[ilayer] + ineuron*(nin_this_layer+1) ;      // Ditto grad
      for (ivar=0 ; ivar<nin_this_layer ; ivar++) {               // Do not include bias
        penalty += wptr[ivar] * wptr[ivar] ;                      // Equation 2-18
        gptr[ivar] -= 2.0 * wpen * wptr[ivar] ;                   // Equation 2-19
        }
      }
    nin_this_layer = nhid_all[ilayer] ;
    }
```

```
  for (ineuron=0 ; ineuron<ntarg ; ineuron++) {
    wptr = final_layer_weights + ineuron * n_final_layer_weights ;
    gptr = grad_ptr[n_all-1] + ineuron * n_final_lay er_weights ;
    for (ivar=0 ; ivar<nin_this_layer ; ivar++) {          // Do not include bias
      penalty += wptr[ivar] * wptr[ivar] ;                 // Equation 2-18
      gptr[ivar] -= 2.0 * wpen * wptr[ivar] ;              // Equation 2-19
      }
    }
  penalty *= wpen ;
  return mse + penalty ;
}
```

The entire first page of this code sets up grad_ptr to contain pointers to the gradient for each layer. There are slightly easier ways to do this, but the algorithm shown here is easy to understand.

The next block of code deals with a subtle but potentially deadly problem. Perfectly reasonable problems can have millions of optimizable weights. When a gradient this long is multiplied by the number of cases in a batch, which also may be large, the product can easily overflow four-byte integers, which are treated as signed in all of my CUDA code. This can also impact memory allocation size. Therefore, we divide the maximum positive number by the number of bytes required to hold the gradient to find the maximum legal number of cases in a batch. Also, current CUDA devices impose an upper limit of 65,535 on the grid dimension that my CUDA code uses to index the case within a batch, so we also impose this upper limit. This determines the minimum number of batches needed. The program should allow the user to increase this value in the unlikely situation that the application is so large and slow that WDDM timeouts occur.

The global variable mlfn_cuda_initialized flags whether the device has been initialized, which involves doing things like allocating scratch memory. A key parameter for the initialization routine is the maximum size of any batch. Given the number of batches just computed (and possibly increased by the user with code not shown here), a simple loop runs through all of the batches that will soon run and keeps track of the largest. We can then call the initialization routine with this number. The calling parameter list is not entirely shown, as most of its components are application-specific and would confuse this presentation.

The global variable cuda_weights_changed flags whether the weights have changed since the last time a CUDA routine was called. If so, we copy the revised weights to the device. As with the initialization routine, the calling parameter list is omitted for clarity.

Gradient computation begins by setting the entire gradient vector to zero. The gradient will be summed in this vector with each batch.

The batch loop is straightforward. For each batch we divide the number of cases remaining to be processed by the number of batches remaining. Add this batch size to the starting index istart to get the stopping index, istop.

The forward pass, in which all activations for this batch are computed and saved, is accomplished by first doing the n_all-1 hidden layers and then the output layer. If this is a classifier, apply the SoftMax modification to the output activations.

The backward pass reverses the order of processing. The output delta is computed and then the output gradient. The hidden layers other than the first are processed, starting with the last and working backward. Finally, the first hidden layer is processed, which is the one fed by the inputs to the model.

The last step in the batch loop is to fetch the gradient, cumulating the contribution of this batch into the grand sum in grad. The number of training cases done so far is updated per this batch, and the starting index for the next batch is set to the stopping index for the batch just done.

After all batches are complete, the summed gradient is divided by both the number of cases and the number of targets because the optimization criterion is also divided by these quantities.

If this is a classifier, we compute the negative log likelihood as the optimization criterion. Otherwise, we compute the mean squared error.

Finally, we implement the weight penalty. The first block of code handles the hidden layers, and the second block does the output layer. These lines of code look complicated, but really they are just straightforward walks through the weights and gradient vector, implementing Equations 2-18 and 2-19 along the way.

Basic Training Algorithms

Now that we know how to compute a good measure of the quality of the model, the mean squared error or log likelihood, and how to compute the partial derivative of the criterion with respect to every weight, we can talk about finding parameters that optimize the criterion. The world is filled with excellent treatments of training algorithms for multiple-layer feedforward networks, so we will not overly dwell on the subject. However, several important algorithms should be discussed to a reasonable degree.

Simulated Annealing for Starting Weights

The common wisdom says that one can choose small random weights for the model and commence training from that point. There is something to be said for this approach; evaluating the optimization criterion is expensive in most real-world problems, and we don't want to waste a lot of computational resources randomly casting about for a good starting point. And it is universally agreed that the point of diminishing returns is reached quickly if someone does randomly search for good starting weights. But it has been my experience that a small amount of *intelligently guided* random search for a starting point is almost always worthwhile.

My favorite starting-weight algorithm is primitive but effective simulated annealing. The idea is that we randomly cast about the parameter space, looking for an area of good performance (minimum criterion). But as time passes we gradually nudge the search location toward areas that have proved good, and we simultaneously lessen the degree of randomness. We will thereby become more and more narrowly focused on a promising region of the parameter space.

The user must specify two search criteria: the initial range for the randomly generated trial weights and the number of trial iterations. I generally use 50 to 1000, with several hundred being a good compromise.

Here is the basic simulated annealing code:

```
anneal_rng = TrainParams.anneal_rng ;        // Initial range for trial weights
anneal_frac = 0.3 / TrainParams.anneal_iters ;   // Anneal_iters is number of tries

for (i=0 ; i<n_weights_to_optimize ; i++)
   center_wts[i] = 0.0 ;

for (i_anneal=0 ; i_anneal<TrainParams.anneal_iters ; i_anneal++) {

   if (i_anneal % 10 == 1)
      factor = anneal_rng * 10.0 ;
   else if (i_anneal % 10 == 2)
      factor = anneal_rng * 4.0 ;
   else if (i_anneal % 10 == 3)
      factor = anneal_rng / 10.0 ;
   else if (i_anneal % 10 == 4)
      factor = anneal_rng / 4.0 ;
```

```
        else
          factor = anneal_rng ;

        for (i=0 ; i<n_weights_to_optimize ; i++)
          wts[i] = center_wts[i] + factor * (2.0 * unifrand () - 1.0) ;
        wvec_to_weights ( n_weights_to_optimize , wts ) ;

        if (ok_to_use_svd)
          find_final_weights ( ... ) ; // Use SVD to find optimal output weights

        crit = trial_error ( ... ) ;

        if (i_anneal==0 || crit < best_crit) {
          best_crit = crit ;
          for (i=0 ; i<n_weights_to_optimize ; i++)
            best_wts[i] = wts[i] ;
          }

        if (i_anneal < 100)
          continue ;

        // Slowly update center and range
        for (i=0 ; i<n_weights_to_optimize ; i++)
          center_wts[i] = (1.0 - anneal_frac) * center_wts[i] + anneal_frac * best_wts[i] ;
        anneal_rng *= (1.0 - anneal_frac) ;
        } // Annealing loop
```

We initialize the center about which weights will be perturbed to zero and then commence the trial loop, which will be done a user-specified number of times.

A modification to the traditional algorithm that I have found to be effective for neural network use is to occasionally use exceptionally large or small perturbations. This is because the user may fail to specify a good perturbation range, either through ignorance or through carelessness. It helps if the algorithm can bail out such users. The block of *if* statements at the start of the loop does just that.

A set of trial weights is derived by randomly perturbing around the center, and then these trial weights are copied to the model's private weights. The call to find_final_weights() uses an extremely efficient algorithm to explicitly compute the optimal output weights given the randomly generated hidden-layer weights. This subject will be discussed in the next section. Then the performance criterion is evaluated by calling trial_error().

If we just set a new record for criterion quality, these weights are saved in best_weights.

If we have done only a few iterations (100 is an ad hoc choice), we just keep randomly casting about. But if we've had a good chance to find a reasonably good region of the parameter space, then we can begin slowly moving your search toward that good region and simultaneously shrinking the radius of the search.

The choice of 100 for the point at which to begin shifting the search region and the choice of 0.3/anneal_iters for the rate at which the shifting takes place were both arbitrary. The reader should feel free to seek better values.

Singular Value Decomposition for Optimal Output Weights

If the output layer is linearly connected to the last hidden layer, which would be the case if we are not doing SoftMax modification, we can often use a technique that tremendously speeds convergence. The process of mapping the activations of the last hidden layer to one or more output targets to minimize mean squared error is just ordinary linear regression. So, why would we use expensive random search such as simulated annealing to find good output weights when we can explicitly compute the exact optimal values? Granted, using regression to find these weights is somewhat expensive itself. But the cost is small when one consider that what it buys is perfectly optimal output weights, given any trial set of hidden-layer weights.

There is little controversy about the wisdom of using linear regression to find the output weights when one is using simulated annealing (or some other random process) to find a starting point for more sophisticated training. The number of parameters that must be searched is reduced, and you are guaranteed that no matter what hidden-layer weights are employed, the output weights will be optimal. It's a no-brainer. But what about using regression during more sophisticated training algorithms, such as the method of conjugate gradients discussed in the next section? The output weight gradient vector for weights computed by regression is identically zero, so it can be ignored. This can significantly reduce the size of the optimization problem, which is good. But regression is not free. The trade-off is whether the time saved for the training algorithm justifies the increased cost of regression. This is a gray area, and my own choice is to trust the training algorithm and avoid regression. Others may differ.

Traditional linear regression involves inverting a matrix. Unfortunately, in neural network applications, this network will often be singular because of the high correlation in the input variables or their resulting activations as they propagate through the network. This is a total deal killer. Thus, we must use an alternative method that is immune to collinearity. It is called *singular value decomposition*, and its code (SVDCMP. CPP) can be downloaded for free from the author's web site. The code contains detailed comments describing its correct use. We'll present a rough overview here.

To use singular value decomposition, one must first create a new SingularValueDecomp object using the following parameters:

```
SingularValueDecomp (
  int nr ,        // Number of rows (cases)
  int nc ,        // Number of columns (neurons in last hidden layer, plus 1 for bias)
  int save_a ) // Preserve the input data matrix for later reference?
```

The following code, shown in outline form, fills in the data and then computes the singular value decomposition. Once this is done, call backsub() for each output to compute the optimal weights for that output.

```
aptr = sptr->a ;    // The data matrix, ncases by nvars (plus 1 for bias) goes here

for (icase=0 ; icase<nc ; icase++) {                    // For each case in the training set
    for (ilayer=0 ; ilayer<nlayers-1 ; ilayer++) {      // For each of the hidden layers
        for (i=0 ; i<nhid[ilayer] ; i++)
            // Compute activation of this neuron from prior layer
        } // For ilayer (each hidden layer)

    // We now have the activations of the final hidden layer
    for (i=0 ; i<nhid_final_layer ; i++)
        *aptr++ = this_layer[i] ; // Move the final hidden layer activations to the SVD input
    *aptr++ = 1.0 ; // Constant term for bias
    } // For all cases

sptr->svdcmp () ;

for (itarg=0 ; itarg<ntarg ; itarg++) {
    bptr = sptr->b ; // True values go here
    for (icase=0 ; icase<nc ; icase++)
        *bptr++ = targets[icase*ntarg+itarg] ;
    sptr->backsub ( 1.e-2 , final_layer_weights+itarg*n_final_layer_weights ) ;
    }
```

One thing to keep in mind is that execution time for SVD blows up rapidly. Hence, for huge problems (very large number of cases and neurons in the last hidden layer), the time can be impractical. Also, in rare pathological cases the algorithm can become numerically unstable. Thus, programmers should allow users the option of disabling SVD.

Stochastic Gradient Descent

The method roughly outlined in this section is old and not used much anymore, so we will only sketch the general idea. But it is simple and easy to explain, so it serves as a good foundation for the more sophisticated algorithm shown in the next section.

Recall that the gradient of a function is the direction of maximum increase of the function. So, by definition, if we move a tiny amount in the opposite direction, we will see, at least locally, a maximum decrease in the function. That's the essence of this primitive but venerable training algorithm. We generate a random set of starting weights and keep moving opposite the gradient. The earliest versions did this one case at a time; the gradient for a single training case would be evaluated, and we would take a tiny step in the opposite direction. Then a new case would be chosen, and we would do the same thing. A relative of exponential smoothing of the gradient, called *momentum*, could be used to prevent wild gyrations in the direction of parameter movement.

At the other extreme, one could evaluate the gradient for all cases and find the average across the entire training set. This would give the most accurate estimate of the gradient, and therefore this approach would usually require the fewest number of iterations to converge to a minimum. Unfortunately, evaluating the average gradient for the entire training set is a time-consuming operation.

Over the years a compromise was reached. The generally best approach when using this (primitive) algorithm is to divide the training set into a large number of small, randomly selected batches. Evaluate the mean gradient for a single batch and adjust the weights accordingly. Then do the same for another small batch. Repeat until convergence is obtained.

This works relatively well because the trade-off between time per batch and number of batches is not equally balanced; they do not cancel perfectly. In particular, when we use more batches, the time per batch usually decreases more rapidly than the number of batches increases.

For example, suppose we divide the training set into 100 batches. The gradient for each batch will have high error variance, thus requiring more adjustments than if we evaluated the gradient for the entire training set. But the number of adjustments needed will not go up as fast as the gradient evaluation time goes down. In this example, which uses 100 batches, the time per batch will be about 100 times smaller than the time to evaluate the mean gradient for the entire training set, but we may need only 50 times more batches. (I pulled that 50 out of the air.) So, overall, we would be ahead by a factor of two.

This training algorithm, though delightfully simple and widely used in the early days of neural networks, suffers from two serious flaws.

- It's difficult to know how far to step in the negative gradient direction each time. If we step too little, an enormous number of steps will be needed to achieve convergence, as shown in Figure 2-4. But if we step too far, we'll overshoot and likely end up increasing the criterion instead of decreasing it. And even if we don't completely overshoot, we may bounce too far and gyrate badly, as shown in Figure 2-5.

- The contours of the criterion as a function of the parameters rarely look like a rubber sheet with a ball weighing down the middle. In fact, Figures 2-4 and 2-5 are quite generous in their portrayal of the difficulty of this optimization problem. Rather, the contours typically resemble the Grand Canyon, a steep, narrow, winding channel cut through a broad plain. The gradient by definition is perpendicular to the "walls" of the canyon, so a simple gradient descent is liable to look more like a zigzag path bouncing wildly back and forth against the walls. Smoothing the path via a momentum modification helps, but rarely just the right amount.

In the next section we will explore a vastly superior training method.

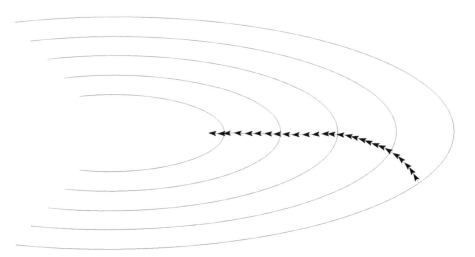

Figure 2-4. *The gradient descent step size is too small*

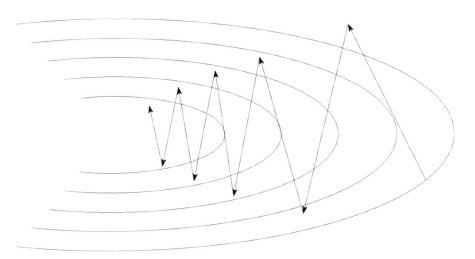

Figure 2-5. *The gradient descent step size is too large*

Conjugate Gradient Optimization

This section will present one of the most popular and effective families of multivariate optimization algorithms. Implementations of it are widely available, both as source code and as linkable libraries, so I won't delve into the code here. But because this is such an important neural network training algorithm, I feel it's worthwhile to at least roughly describe the algorithm so that its users will have some understanding of its operation. This section does involve some mathematics beyond the level elsewhere in this book, and it may safely be skipped.

There is an almost magical beauty in conjugate gradient methods. They have nearly (or entirely) the power and convergence speed of full second-order methods. Yet they require neither the computation nor even the storage of any second-order information. That is a matter of no small importance. When there are more than a handful of variables to be optimized, storage and manipulation of an n^2 Hessian matrix can be a heavy burden. And computing all those second derivatives is prohibitively expensive. An algorithm that behaves as if it were based on second-order information yet that neither explicitly computes nor stores anything having to do with second derivatives is one step this side of a miracle.

The full mathematics of conjugate gradients is not overly complicated, but it is rather long and tedious. Therefore, our presentation will be cursory and will feature an intuitive approach. Readers who want the complete story, should start by carefully studying the excellent general discussions of multivariate minimization in numerous standard texts.

A major goal of intelligent optimization algorithms is the elimination of the bouncing off the walls that plagues naive algorithms. We mentioned that introducing momentum helps but does not solve the problem entirely. Explicit second-order methods do much better than momentum by using second derivatives (or their approximations) to rotate the gradient so that it points toward the minimum, and they also provide a good estimate of how far to step, a crucial bit of information for minimizing the number of steps. Unfortunately, in most practical situations, those methods are so uneconomical (storage space and derivative computation time) that they simply cannot be used, no matter how good their theoretical performance might be.

The root cause of the difficulties associated with naive gradient descent is that every time we descend in some direction, we lose a sizable amount of your hard-earned progress from previous steps. We compute the gradient at the starting point and step in that direction. The gradient at that new point is perpendicular to the first search direction if we have done a good job of choosing a jump distance. We then step in that new direction. Unfortunately, when we get there, we are dismayed to find that the gradient is not much different from what it was at the starting point. We already put a tremendous amount of effort into computing that gradient and stepping in that same direction not too long ago, and now we must do it again. Ouch! Wouldn't it be nice if we could instead move along a direction that had the property of not taking back progress already made? We can. All we do is minimize in a direction that is *conjugate* to the previous direction (or directions).

Most readers know what it means for two directions (vectors) to be perpendicular (orthogonal). In words, it means that motion along one of them does not generate any motion along the other. In mathematics, it means that their dot product is zero.

What does it mean for a pair of vectors to be conjugate? The concepts are not entirely different. A major difference is that the concept of orthogonality involves only the pair of vectors in question. The concept of conjugacy additionally involves a quadratic function (a multivariate polynomial having no terms higher than the second order). In other words, we cannot really say that two vectors are simply conjugate. To be precise, we must say that two vectors are conjugate *with respect to some quadratic function*. In practice, the quadratic function is assumed to be known, so we usually omit any mention of it. But the reader must understand that its presence is crucial. We can now state what is implied by two vectors, say *r* and *s*, being conjugate. It means that when we move along one of them, say *r*, the *change* in the gradient of the function is perpendicular to the other vector, *s* in this case. An equivalent mathematical formulation of this condition is shown without proof in Equation 2-20. See any of the many available references for proof. In that equation, *H* is the Hessian matrix of second derivatives of the function.

$$r'H \ s = 0 \qquad\qquad (2\text{-}20)$$

What are the implications of this to function minimization? First, it should be emphasized that the definition of conjugacy depends on the function being quadratic, which implies a constant Hessian matrix. In real life we will never be blessed in this way, so everything we are about to say will be compromised to the degree that the function is not quadratic. On the other hand, many practical functions have a fairly constant Hessian, especially in the neighborhood of a minimum. So, we generally are quite safe in this assumption.

With that warning taken, let's think about minimization. We start at a point, compute the gradient there, and step in the negative gradient direction. If we are ambitious, we may even descend to the minimum along that line using an efficient algorithm, an operation that requires only computation of the criterion but not the gradient. The naive approach would be to compute the gradient again (if we are at the minimum in that direction), noting that it will be perpendicular to the previous search direction and descend once more.

But what if we instead computed a direction that is conjugate to the first search direction and minimized along that line? By the definition of conjugacy, any change in the gradient resulting from that minimization motion will be orthogonal to the previous search direction. In other words, if we were already at a minimum in that previous search direction, we will remain at a minimum in that direction. This new minimization does not cost us any of our winnings in the previous minimization. That is in sharp contrast to the situation with naive steepest descent, in which each step in a negative gradient direction costs us much of the progress already made in previous directions.

It's time for an example. Look at Figure 2-6 and Equation 2-21. A specific case involving explicit points and search directions is illustrated there.

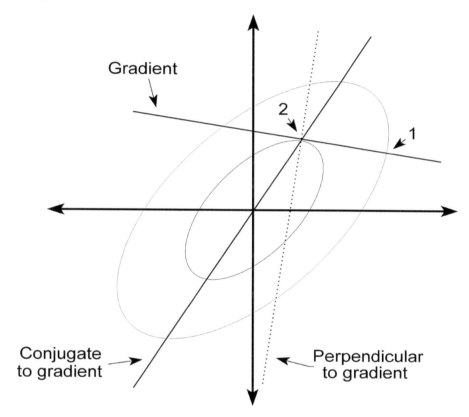

Figure 2-6. *Conjugate gradient optimization in two dimensions*

$$f(x,y) = x^2 + y^2 - 1.5xy$$

$$\frac{\partial f}{\partial x} = 2x - 1.5y$$

$$\frac{\partial f}{\partial y} = 2y - 1.5x$$

$$\frac{\partial^2 f}{\partial x^2} = \frac{\partial^2 f}{\partial y^2} = 2$$

$$\frac{\partial^2 f}{\partial x\, \partial y} = -1.5$$

(2-21)

This example is primitive in that it involves only two independent variables, and the function being minimized is a perfect quadratic. However, it is instructive to study its geometry and actually work through some steps. The reader will then ideally better appreciate the power of the conjugate gradient method. Observe that Equation 2-21 not only tells us the form of the function but also shows us how to compute the gradient at any point. Also notice that the Hessian is constant, with the value 2.0 at both diagonals and -1.5 in the off-diagonal positions. The following discussion is terse, but most readers should be able to verify each step. In fact, that is recommended.

We have a simple quadratic function for which we seek the minimum. The search starts at point 1 whose location happens to be $(6, 4)$, an arbitrarily chosen point. The gradient at that point is $(6, -1)$. A line in that direction through point 1 passes through the inner of the two level ellipses that are shown. Note that, as expected, the gradient is perpendicular to the ellipse at the point of tangency. If we parameterize that line as point 1 plus t times the gradient, the derivative of the parameterized function with respect to t is $92\,t + 37$. Set that equal to zero to find the location of the function's minimum along the gradient line. The solution, $t = -0.4022$, gives us a location of approximately $(3.587, 4.402)$. That line minimum is labeled as point 2.

Everything has been straightforward so far. Here is the important part: if we simply compute the gradient at this new point and naively minimize along that direction, we would be following the dotted line. That's not terrible, but it's not very good either. Let's compute a conjugate direction instead. (Since we have only two variables, there is only one direction that is conjugate here. In higher dimensions there will be more.) We compute the conjugate direction using Equation 2-20. Let r be the gradient direction and H be the Hessian. If $s = (x, y)$, then we get $13.5x + 11y = 0$. Notice in Figure 2-6 that the line through point 2 in this direction passes directly through the minimum of the function! Lovely!

Actually, that observation is not unexpected. In fact, it can be proved that if we are minimizing a perfect quadratic in n variables, then n line minimizations in directions that are mutually conjugate are all that are needed to find the minimum. This should agree with intuition if we recall what it means for directions to be conjugate. Every time we minimize along one of these directions, previous minimizations are preserved. After we have done this n times, we are at a minimum in all directions. As long as the directions span the entire space, which is easily proved, we can do no better.

This is all very nice, but so far we have really done nothing special. When we computed the direction that is conjugate to the previous search direction, we casually glossed over the role that the Hessian played. Since this was a simple contrived example, we knew the Hessian. Unfortunately, in real-life problems, the Hessian will be difficult or impossible to compute. In those rare cases in which it can be economically computed, we might as well use Newton's method or a relative. Here is where the magic enters.

Let $\mathbf{g}_i$ be the negative gradient at point i in the search for the minimum. Let $\mathbf{h}_i$ be the direction that we will search from this point. To initialize the algorithm, set $\mathbf{g}_0$ and $\mathbf{h}_0$ both equal to the negative gradient at the starting point. At each step we must keep track of the gradient and the search direction at the previous step (but only that one previous step). Compute the search direction for the current step using Equation 2-22.

$$\gamma = \frac{(\mathbf{g}_i - \mathbf{g}_{i-1}) \cdot \mathbf{g}_i}{\mathbf{g}_{i-1} \cdot \mathbf{g}_{i-1}}$$

$$\mathbf{h}_i = \mathbf{g}_i + \gamma \mathbf{h}_{i-1}$$

(2-22)

It can be shown that as long as we are careful to locate the minimum in each successive search direction, the sequence of $\mathbf{h}_i$ search directions will be mutually conjugate to the extent that the Hessian is constant. The remarkability of this result cannot be overstated. We have attained that holy grail of multivariate optimization, a set of mutually conjugate search directions, with no explicit knowledge of the Hessian. In fact, unlike some other methods that also avoid computing the Hessian (such as Levenberg-Marquardt), we don't even have to store any approximation to it. The only extra storage space that we need is a few vectors to preserve the gradient and the search direction from the previous step, a mere pittance. I consider this algorithm to be one of the seven wonders of modern mathematics.

Astute readers will immediately notice the similarity of this algorithm to traditional backpropagation with momentum. The second line in Equation 2-22 tells us that the new search direction is the gradient at this point plus a scalar times the previous search direction. There are only two differences. First, in traditional backpropagation with momentum, the step size is set according to a preordained schedule, while in the conjugate gradient algorithm, the step size is carefully chosen to be the line minimum along the search direction. Also, in traditional backpropagation, the momentum term is fixed or slowly varying according to a schedule. In the conjugate gradient algorithm, the momentum is optimally adjusted for each new step. These are actually two quite significant differences, and they account for the typically huge difference in performance.

CHAPTER 3

Restricted Boltzmann Machines

This chapter explores what is probably the most common building block of deep belief nets: the restricted Boltzmann machine (RBM). There are numerous excellent treatments of RBMs; my favorite "introduction" is *Learning Deep Architecture for AI* by Yoshua Bengio (Now Publishers, 2009) because the discussion, though relatively short, is enclosed in fabulous background and supplementary material. When it comes to practical aspects of training RBMs, "A Practical Guide to Training Restricted Boltzmann Machines" by Geoffrey Hinton (2010) can't be beat. Also, the numerous technical papers by Geoffrey Hinton cover specific aspects of RMBs in glorious detail. Finally, deeplearning.net is an incredible resource. Because of this wealth of material, we will avoid unnecessary duplication. This chapter will be limited to an outline of the essentials of RBMs, in other words, the information necessary to understand and use the programs presented here.

What Is a Restricted Boltzmann Machine?

An RBM is in essence a two-layer neural network that is bidirectional and partially controlled by randomness. Also, the RBMs studied here have strictly binary activations of their hidden neurons, and binary inputs are favored. Continuous inputs and hidden activations are much more complex and cantankerous. Their study is still in the early stages.

An RBM does not have an output layer as such. It has only an input layer (usually called the *visible* layer in the context of RBMs) and a hidden layer whose activations often serve as outputs that will be used in ways described later.

© Timothy Masters 2018
T. Masters, *Deep Belief Nets in C++ and CUDA C: Volume 1*, https://doi.org/10.1007/978-1-4842-3591-1_3

Recall that for a traditional feedforward network, the activation of a neuron in a given layer is computed by forming the dot product of a weight vector with the activations of the prior layer, adding a bias term, and applying the logistic activation function. The same formulation applies to an RBM but with one exception. Instead of this procedure determining the activation of a hidden neuron given the inputs, it determines the *probability* that the hidden neuron will have the value one (as opposed to zero). Then the hidden neuron is randomly set to either zero or one according to this probability. This is made explicit in Equation 3-1.

We have a weight matrix W that has D columns, one for each input (visible neuron), and as many rows as there are neurons in the hidden layer. The bias c and input (visible neurons) v are column vectors. Let $W_{i\bullet}$ signify row i of the weight matrix, and let c_i be element i of the hidden neuron bias vector. Then the probability that hidden neuron i is one (versus zero) is given by Equation 3-1, in which $f(.)$ is the logistic function shown in Equation 1-4.

$$P\left(h_i = 1 \mid v\right) = f\left(c_i + W_{i\bullet}v\right) \tag{3-1}$$

Aside from their binary and stochastic natures, what differentiates RBMs from traditional feedforward networks is the fact that they are bidirectional. We just saw that, given an input (visible neuron) vector, we could compute the probability that each hidden neuron is one. Let $W_{\bullet j}$ signify column j of the weight matrix, and let b_j be element j of the visible neuron bias vector. Then for some set of hidden neuron states h, the probability that visible neuron j is one (versus zero) is given by Equation 3-2.

$$\mathbf{P}\left(v_j = 1 \mid h\right) = f\left(b_i + W'_{\bullet j}h\right) \tag{3-2}$$

Here's where it gets really interesting. Suppose we set the visible neurons v to some random state and use Equation 3-1 to set the hidden neurons to random states with the computed probabilities. Then we use Equation 3-2 to randomly set the states of the visible neurons. Apply these "bounced back" inputs to Equation 3-1 to get new hidden-layer activations. Repeat many times. (We could just as well have begun by setting the hidden neurons to random initial states and used Equation 3-2 as the first step. The final result would be the same.) After a considerable number of back-and-forths, the sequence of states will converge to a stable Markov chain. The rate at which convergence is obtained is called the *mixing rate* and depends on W (smaller weights lead to faster mixing), but it is generally fast enough to be practical. Once convergence to a stable chain is obtained (which we cannot verify but which in practice we can count on happening after a reasonable number of iterations), the set of visible and hidden neuron activations obtained from continued cycles represents random samples from a probability distribution that depends on the weight matrix and bias vectors. They will not be independent, of course, but they will be unbiased in that they are honestly from the probability distribution implied by the weights and biases.

So, what's the big deal? We specify a parameter set consisting of a weight matrix, hidden neuron bias vector, and visible neuron bias vector. In return we get a means of generating random samples of visible and hidden neuron states that have a fixed probability distribution. So what?

The big deal is that we have collected a training set, a (presumably) random sample of visible states taken from some universe with a fixed probability distribution. What if we could find a parameter set such that the distribution of random visible neuron states produced by the Markov chain of alternating Equations 3-1 and 3-2 mimics the distribution of cases in the training set? We would then have a representation of the structure of the training data. The RBM would encapsulate the patterns inherent in the data. In particular, Equations 3-1 and 3-2 would define an explicit mapping between input vectors and hidden neuron activations *that is based on the patterns in the data*. Thus, the hidden neurons present an alternative and often (usually!) clearer representation of the information in the inputs. Moreover, if there are fewer hidden neurons than inputs, the representation is more compact.

Reconstruction Error

Suppose we are able to find a parameter set that provides a good (in some sense) bidirectional mapping between input vectors and hidden neuron activation vectors. Then it is reasonable to assume that when we use Equation 3-1 to map an input vector to a hidden neuron activation vector and then use Equation 3-2 to bounce activations back to the visible layer, the visible activations attained will resemble the original input. That's the whole idea of alternate representations of information. For this reason, we often want to compute a measure of this agreement.

Using these equations to define probabilities and then sampling to set the activations to zero or one is acceptable. However, this sampling introduces annoying randomness into what might rather be a deterministic measure. For this reason, we are inclined to stick with probabilities in both directions, as shown in Equations 3-3 and 3-4, which define the entire hidden and visible probability vectors. The logistic function $f(.)$ is applied element-wise.

Later we will use h and v strictly for activations, but for the moment we allow them to also represent probabilities to simplify the presentation of this section.

$$h = f\left(c + Wv\right) \qquad (3\text{-}3)$$

$$v = f\left(b + W'h\right) \qquad (3\text{-}4)$$

Let v_i denote the ith element of the reconstruction v, and let x_i similarly denote the ith element of the input x. Then the most common measure of reproduction error is the mean squared error shown in Equation 3-5. If the inputs are either binary or can be considered to be binomial probabilities, then the cross entropy shown in Equation 3-6 may be preferred in that it is more sensitive to extremely erroneous reconstructions. (Consider what happens if v_i is extremely close to zero or one when x_i is the opposite.) In both cases, this quantity would be averaged over the training set.

$$ReconErr = \sum_i \left(v_i - x_i\right)^2 \qquad (3\text{-}5)$$

$$ReconErr = -\sum_i \left[x_i \log\left(v_i\right) + \left(1 - x_i\right)\log\left(1 - v_i\right)\right] \qquad (3\text{-}6)$$

Intuition would lead one to believe that either of these might be a great criterion to optimize to find a desirable parameter set. This intuition would be wrong. If the goal is to reproduce the input data in a training set, these are fabulous criteria, but that is not the usual goal! A simple identity transform, which does not find an alternative representation at all, would have perfect reconstruction. The real goal is to encapsulate legitimate patterns that will also appear out-of-sample. For this task, a different optimization criterion is needed, although a reconstruction error is a useful quantity to monitor as learning progresses. It is not without merit, just without supremacy.

Maximum Likelihood Training, Sort Of

Much of this section, especially the beginning, is intensely theoretical and may safely be skipped by mathematically challenged readers. None of this initial development is used in programming RBMs, nor is it crucial to understanding RBM algorithms. It is included only to let interested readers know why in practice we cannot do what we would like to do. Life is hard, and this is one of the reasons why (for some of us anyway).

Glance back at Equation 1-3. In that equation we showed how the bias could be conveniently absorbed into the weight matrix. This is almost always inconvenient for RBMs; we will mostly keep the hidden and visible bias as vectors separate from the weight matrix, especially in programs. However, for this brief presentation it is most convenient to absorb the bias vectors into the weight matrix by implicitly including a single visible neuron and a single hidden neuron with the constant value 1.0 so that their weights will be the corresponding biases.

With this in mind, suppose we have an RBM model with weight matrix W. Define the scalar *energy* of a visible and hidden neuron state for this model, as shown in Equation 3-7. For any given W, which completely defines the RBM model, the probability of a visible and hidden neuron state is given by Equation 3-8, where $Z(W)$, shown in Equation 3-9, is the normalizing term that ensures that the probabilities sum to one.

$$E(v,h,W) = -h'Wv \tag{3-7}$$

$$P(v,h\,|\,W) = \frac{1}{Z(W)}\exp\big(-E(v,h,W)\big) \tag{3-8}$$

$$Z(W) = \sum_v \sum_h P(v,h\,|\,W) \tag{3-9}$$

We get the first hint that this line of inquiry will not end well when we ponder Equation 3-9. Those two summations are over every possible visible and hidden state, so the number of terms being summed is two to the power of the total number of neurons, visible plus hidden. Yikes!

Recall that on page 94 we pointed out that a reconstruction error, while interesting and useful for observation, is not ideal for finding an optimal RBM model. The ultimate goal is to find a model that encapsulates authentic patterns in the training data, rather than just spitting back inputs. Such a model is thereby able to reproduce the distribution of the training data (and hence the universe, we hope), so we are inspired to find a model that maximizes the likelihood of the training set. In other words, we seek a maximum likelihood weight matrix W, a weight matrix that enjoys, from among all possible weight matrices, the highest probability of producing the training set that we observed.

Despite any apprehension based on the frightening Equation 3-9, let's plug on toward the goal of finding a maximum likelihood W. The log likelihood of a single training case x is found by summing Equation 3-8 across the marginal of all possible hidden neuron states and then taking the log, as shown in Equation 3-10. As we did on page 16 for SoftMax maximum likelihood, instead of using the straight likelihood, which is multiplicative, we use its log. This way we can find the (log) likelihood of the entire training set by summing the log likelihood of the individual cases.

$$\log P(x|W) = \log \sum_h \exp(-E(x,h,W)) - \log Z(W) \tag{3-10}$$

The first term in Equation 3-10 is not as bad as Equation 3-9 because we are summing just hidden states, not hidden and visible both. But it is nonetheless intractable in all but the smallest models. Still, this is a handy equation to have in our toolbox because in small "research" problems it is easy to evaluate it using Equations 3-7 and 3-9 for the intermediate terms.

Equation 3-11 shows the partial derivative of Equation 3-10 with respect to an individual element of W. Equation 3-12 is a more general expression of the partial derivative, appropriate for our use, which involves random sampling to approximate expectations rather than explicit calculation. In Equation 3-12, the angle brackets denote statistical expectation under the named distribution.

$$\frac{\partial \log P(x|W)}{\partial w_{ij}} = \sum_h x_j h_i P(h|x,W) - \sum_v \sum_h v_j h_i P(v,h|W) \tag{3-11}$$

$$\frac{\partial \log P(v|W)}{\partial w_{ij}} = <h_i v_j>_{W,TrainingSet} - <h_i v_j>_W \tag{3-12}$$

The first term, not surprisingly called the *positive term*, is easy to handle. Just pick a case from the training set. Then to get an unbiased sample of h given W and the chosen training case, apply Equation 3-1. You either can sample h_i as zero or one using the probability given by that equation or can use the probability directly. Later, when the gradient algorithm is presented, we'll see when one must sample 0/1 and when one would preferably use the probability.

Unfortunately, the second term in Equation 3-11, called the *negative term*, is intractable in any practical application, as the number of summands going into the double sum is two to the power of the total number of neurons, visible and hidden. The best we can do is view the situation through the lens of Equation 3-12 and attempt to get a reasonably unbiased and low variance sample of this term from its distribution. We then hope that the average across a large number of samples will be close to the true expectation.

The algorithm for doing so, while tractable, is quite expensive. Fortunately, there is a much faster algorithm that almost always provides an inferior but acceptable approximation. We'll present these algorithms in the next section.

Contrastive Divergence

It must be emphasized that our approach to training an RBM is not a nice deterministic algorithm, one that explicitly calculates the exact gradient like we could do when training a multiple-layer feedforward network. Rather, we will do *stochastic gradient descent*, an approach in which we take random samples of the positive term, which depends on training cases, and take other random samples from the RBM's distribution implied by W, which (ideally, though not quite in practice) is independent of training cases. We then subtract them as shown in Equation 3-12 and hope that errors induced by randomness will largely cancel. They almost always do.

A few paragraphs back we saw that the positive term is easy. But how do we get samples from the theoretical distribution of the RBM as determined by W? Recall that on page 92 I pointed out that if we bounce back and forth between visible and hidden activations using Equations 3-1 and 3-2 for many repetitions, we will converge toward (correlated) samples from the true RBM distribution.

This convergence will happen for any random starting state, but if the starting state is chosen from a distribution that is close to the RBM's theoretical distribution, then convergence will happen quickly. In fact, if the random starting state happens to be chosen from the exact RBM distribution, convergence will be instantaneous, so no iteration will be needed. We naturally assume that as learning occurs, the distribution of the RBM will approach that of the training set; this is the whole point of our endeavor! So, it behooves us to choose as a starting point the training case x under consideration. Early in training this won't do any good if the initial weights were randomly chosen because then the empirical distribution from which x is drawn will only by good luck bear any resemblance to the starting RBM distribution. But as training progresses, they will become closer and closer, making this starting selection more and more beneficial. And later we will present a method for intelligently choosing an initial weight matrix W that causes the RBM's distribution to be closer to that of x than a random weight matrix would usually be.

We now present the basic algorithm for stochastically estimating the gradient of the log likelihood function at a training case *x*. Recall that when we get the hidden neuron activations via Equation 3-1 or 3-3 and when we get the visible activations via Equation 3-2 or 3-4, we can either sample from 0/1 per the computed probabilities or use the probabilities directly. There are three rules to be followed.

- When getting the hidden activations, for all iterations except the last we *must* sample from 0/1. This is critical to the regularization inherent in RBMs.

- When getting the hidden activation from the final iteration for use in Equation 3-12, it is almost certainly best to use the probability directly.

- When getting the visible activations, either is fine, although directly using the probabilities is considered better by most experts. This is called a *mean field* approximation.

The following notation will be used:

W Weight matrix, a column for each visible neuron and a row for each hidden neuron

b Column vector of visible neuron biases

c Column vector of hidden neuron biases

K The number of Monte Carlo iterations to perform

x The training case being processed (column vector)

q_{Data} Vector of probabilities under the data distribution that each hidden neuron will be one (as opposed to zero)

h_{Data} Hidden neuron activation vector under the data distribution, zero or one

p_{Model} Vector of reconstruction probabilities under the model distribution that each visible neuron will be one (as opposed to zero)

v_{Model} Reconstructed visible neuron activation vector, zero or one

q_{Model} Vector of probabilities under the model distribution that each hidden neuron will be one (as opposed to zero)

h_{Model} Hidden neuron activation vector under the model distribution, zero or one

It is to be understood that *p* is a vector of length equal to the number of inputs (visible neurons), and it contains probabilities computed by Equation 3-2 or 3-4. Each element of *v* is individually sampled from 0/1 according to these probabilities. The hidden neuron probabilities and activations are defined similarly.

$$\boldsymbol{v}_{\text{Data}} = \boldsymbol{x}$$

$$\boldsymbol{q}_{\text{Data}} = f(\boldsymbol{c} + \boldsymbol{W}\boldsymbol{v}_{\text{Data}}) \qquad\qquad \text{Equation 3-3}$$

Optionally compute the reconstruction error using the slow, accurate method.

$$\boldsymbol{q}_{Model} = \boldsymbol{q}_{Data} \qquad\qquad \text{MC chain loop below initializes from data}$$

$$k = 0$$

while $k < K$ K must be at least 1

    Sample $\boldsymbol{h}_{\text{Model}}$ from $\boldsymbol{q}_{\text{Model}}$ This sampling is critical; must not use q

$$\boldsymbol{p}_{\text{Model}} = f(\boldsymbol{b} + \boldsymbol{W}'\boldsymbol{h}_{\text{Model}}) \qquad\qquad \text{Equation 3-4}$$

If k=0, optionally compute the reconstruction error using the fast method.

    if *mean field*

        $\boldsymbol{q}_{\text{Model}} = f(\boldsymbol{c} + \boldsymbol{W}\boldsymbol{p}_{\text{Model}})$

    else

        $\boldsymbol{v}_{\text{Model}}$ is sampled from $\boldsymbol{p}_{\text{Model}}$

        $\boldsymbol{q}_{\text{Model}} = f(\boldsymbol{c} + \boldsymbol{W}\boldsymbol{v}_{\text{Model}})$

    $k = k+1$

end while

if *mean field*

    Visible bias gradient $= \boldsymbol{p}_{Model} - \boldsymbol{v}_{Data}$

    Hidden bias gradient $= \boldsymbol{q}_{Model} - \boldsymbol{q}_{Data}$

    Weight gradient $= \boldsymbol{q}_{Model}\,\boldsymbol{p}'_{Model} - \boldsymbol{q}_{Data}\,\boldsymbol{v}'_{Data}$ This product is a matrix

else

    Visible bias gradient $= \boldsymbol{v}_{Model} - \boldsymbol{v}_{Data}$

    Sample $\boldsymbol{h}_{Data}$ from $\boldsymbol{q}_{Data}$

    Hidden bias gradient $= \boldsymbol{q}_{Model} - \boldsymbol{h}_{Data}$

    Weight gradient $= \boldsymbol{q}_{Model}\,\boldsymbol{v}'_{Model} - \boldsymbol{h}_{Data}\,\boldsymbol{v}'_{Data}$

A few things should be noted about this algorithm. First, the weight gradient is a matrix that, like $\boldsymbol{W}$, has a row for each hidden neuron and a column for each visible neuron. The products given in Equation 3-12 are efficiently represented in the algorithm by showing them as the product of a column vector for hidden neurons times a row vector for visible neurons.

There are two different places in the algorithm in which one can compute the reconstruction error. This error has no use in the training algorithm itself, but it is nice to display it for the user. Regardless of which place we choose, Equations 3-3 and 3-4 are used to jump from the visible layer to the hidden layer and then bounce back to the visible layer. The reconstruction error will compare the original data with the reconstructed data. The only question is whether we use the raw probabilities from these equations or samples based on the probabilities. As was pointed out in that section, using the raw probabilities is more accurate.

The practical problem with using probabilities is that then we need to do a significant amount of computation that is not used anywhere else. For computing the gradient, it is vital that the hidden neurons be sampled. The information loss because of sampling is critical to the robust behavior of the model. But if we compute the reconstruction error from the probabilities, then for each visible neuron we have to sum a dot product over all hidden neurons to evaluate Equation 3-4. This is expensive, a total number of multiplications and additions equal to the number of inputs times the number of hidden neurons. And these unsampled values will have no other use. In a time-critical situation, the usual situation, this is a huge waste. It is much more efficient, if slightly less accurate, to compute the reconstruction error in the first pass through the Monte Carlo chain loop using the sampled hidden activations that are required for the gradient.

Lastly, a surprising fact about gradient computation is that a huge amount of information about the model's distribution is obtained from the *first* pass through the Monte Carlo chain loop ($K=1$). The bias from using just one pass means that more training epochs will be needed and convergence cannot be perfect. But because the chain loop is so expensive, the trade-off is almost always well worthwhile, at least for early epochs.

Weight Penalties

As was the case for supervised training discussed in the prior chapter, large weights can be problematic. We saw that in the context of supervised training, colinearity of predictors was a significant potential source of overly large and problematic weights. This is not a problem with unsupervised training. But the two situations still share the problem of overfitting. If the RBM weights are prevented from becoming large, perhaps the ability of the model to encapsulate the distribution of the training data will be slightly impaired, but at the same time the model will be less likely to include "patterns" that are due to noise. Thus, the model will be better able to generalize to data outside the training set.

An even more important reason for discouraging large weights in an RBM has to do with the mixing rate. We saw on page 92 that the mixing rate is the speed at which the "bouncing back and forth" between the visible and hidden units converges to an unbiased random sample from the RBM's distribution implied by the weights. This rate depends on the size of the weights, with smaller magnitude rates producing faster mixing. Since the degree to which the negative term in Equation 3-12 is an unbiased sample from the RBM's distribution impacts the accuracy of the stochastic gradient estimate, we want the mixing to be as fast as possible. Thus, we want the weights to be as small as possible, consistent with them being large enough to adequately model the data.

A good weight penalty is provided by the same squared magnitude that was used for supervised training. This is given by the penalty shown in Equation 2-18. It has the partial derivative shown in Equation 2-19. *PenFac* is the user's specified penalty factor. Because RBM training does not include the ability to compute a criterion function, only its gradient, the value of the penalty itself is of no use. But we still add the partial derivative to each gradient term computed by the algorithm shown earlier.

Encouraging Sparsity

It should be obvious that a hidden neuron that is always on or always off (a situation usually caused by its bias having excessive magnitude) is a worthless neuron. Therefore, we are inspired to, at a minimum, gently encourage neuron activation rates to avoid extremes.

There is often a second reason for encouraging neurons to not only avoid extremes but gravitate to activation rates (fraction of training cases for which it is turned on) that are small but positive, such as 0.1 or so. Later we will explore several methods to see the patterns learned by neurons. When we do so, it will become clear that these patterns are most interpretable by humans when they activate well under half the time. If some neuron activates for about half of the training cases, its reason for existing may not be obvious. But if it activates for only one-tenth of the training cases, it will probably be more apparent exactly what patterns in the training data they are responding to.

A good penalty function to encourage sparse activation is the cross entropy between the desired (p) and actual (q) activation rates. This function, defined in Equation 3-13, is graphed in Figure 3-1 for a desired rate of 0.1.

$$Spen = -PenFac * \Big[p\log(q) + (1-p)\log(1-q) \Big] \tag{3-13}$$

101

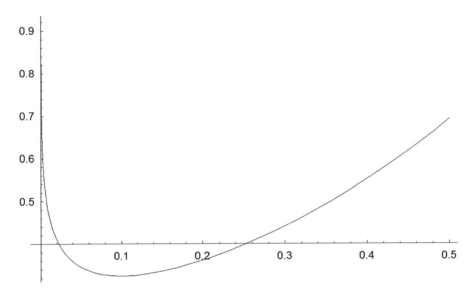

Figure 3-1. *Sparsity cross-entropy penalty function for p=0.1*

In this equation, *PenFac* is a small user-supplied penalty strength factor. This function satisfies the basic need of having its minimum when *p=q*. But three properties make this function especially appealing. First, as the achieved activation rate approaches the limits of zero and one, the log terms blow up, meaning that neurons that are always on or always off become essentially impossible. Second, the function is relatively flat near its minimum, meaning that it will not violently force conformity to the exact desired activation rate but will allow some flexibility as long as the achieved rate is fairly close to the desired rate.

The third property is great for programming the penalty: the cross entropy's derivative with respect to the neuron's net input is delightfully simple, *q–p*. The derivative of the net input with respect to a given weight is just the input (visible neuron activation) feeding through that weight. So, by the chain rule, we have the simple formula in Equation 3-14 for the partial derivative of the sparsity penalty with respect to the weight connecting visible neuron *j* to hidden neuron *i*.

$$\frac{\partial Spen}{\partial w_{ij}} = PenFac * v_j (q_i - p) \tag{3-14}$$

It's not practical (and probably not even desirable) to evaluate this function for each training case. Rather, we let v_j be the mean across all training cases of input *j*. For q_i, we can compute the mean activation of hidden neuron *i* in a batch and then exponentially smooth this quantity to prevent rapid changes that might whipsaw the gradient.

The primary use for a sparsity penalty is to encourage mean activations to be some small but not tiny rate, such as 0.1, to make weight interpretation easier. But I have found it useful to include two additional sparsity penalties in RBM training. One is invoked only when q is tiny, such as less than 0.01, and the other is invoked when q is very large, such as greater than 0.99. By using a large penalty factor, one can further strengthen the force and keep neurons from saturating at zero or one. This is handy if one wishes to use little or no "traditional" sparsity penalty.

Finding Initial Weights

The first step in training an RBM network is to come up with a set of weights and biases that will serve as an effective starting point for optimization using some gradient-based approach. The traditional method seems to be just generating a small random set of weights and then choosing initial bias vectors with a modest degree of intelligence. This may indeed be fine because the early stages of subsequent gradient-descent training rapidly improve the quality of the parameters. Since RBM operations are computationally expensive, it can be argued that any approach more sophisticated than a single initial random shot at starting weights is a waste of resources.

However, it has been my experience that a small amount of searching for a good starting weight set is worthwhile. A stage of computing reconstruction error is much faster than a stage of gradient evaluation, especially if the gradient evaluation employs more than one iteration of the Markov chain. One minute of computer time spent trying a variety of parameter sets can greatly reduce the probability of accidentally starting out in the Bermuda Triangle of weight space. On average it may be a wasted minute, but every now and then a lot of agony is spared.

There are some basic principles to be followed in choosing trial weight sets. These include the following:

> → It's better to have the magnitudes be too small than too large. If they are small, the training algorithm will pull them to better values soon enough. But if they are too large, they may cause some neurons to saturate in an always-on or always-off state, thus effectively destroying the link between visible-hidden pairs and driving the gradient to zero. Remember, your goal is to find starting weights, which are values more likely to be in a "good" region of the parameter space than a "bad" region. We don't need to try to find even remotely optimal weights.

→ There should be equal probability of generating positive and negative weights. To take a trivial example, suppose we have two inputs and there are two patterns: the first input is positive, and the second is negative, versus the first is negative and the second positive. Suppose your trial weight generator produces only positive weights, with the misguided idea that encouraging positive weights produces more interpretable patterns. Since the optimal weight set would have opposite signs for the inputs, no trial weight set could come close to the optimal weight set.

→ The variation of the weights must depend on the dimensions of the weight matrix. If you are summing a dot product of neuron activations and a weight vector, the magnitude of the sum will depend on the length of the vector. In particular, under reasonable assumptions of independence and identical activation distributions, the standard deviation of the sum is proportional to the square root of the number of terms in the dot product. Unfortunately, we have to deal with sums in two directions: across columns for mapping the visible neurons to the hidden and down rows for mapping hidden to visible. If the weight matrix is profoundly nonsquare, we may have a quandary, in which case we should always err on the side of weights that have small magnitude. But in practice the number of hidden neurons is usually well within an order of magnitude of the number of visible neurons, so we have plenty of leeway. A reasonable procedure is to use the harmonic mean of the dimensions, which is the square root of their product. This leads to the guideline in Equation 3-15 for the variation of the weights.

$$Variation \propto \frac{1}{\sqrt{\sqrt{nvis * nhid}}} \tag{3-15}$$

To create a range of variations, I use Equation 3-16 to define a "grand" variation for an entire weight set, and I use Equation 3-17 to sample each weight. Uniform(0,1) is a uniform random number in (0,1).

$$Variation = \frac{4 * Uniform(0,1)}{\sqrt{\sqrt{nvis * nhid}}} \tag{3-16}$$

$$Weight = Variation * \left(Uniform(0,1) - 0.5\right) \tag{3-17}$$

Hidden Neuron Bias

It is vital that no hidden neuron net input be inclined to be so extreme that the neuron is saturated to a probability near zero or one for most or all training cases. This will certainly happen if the bias has an extreme magnitude, which is why a common habit is to initialize all hidden neuron biases to zero. This habit is fine if there are so many weights that their net sum likely sums to around zero relative to their variation. But if there are not a lot of inputs, then near saturation, though unlikely if the weights are small, is possible. All inputs are, by definition, non-negative. So, for example, if a random trial weight set happens to have a row that is severely unbalanced in positive and negative weights, the dot product of this row with the inputs can be far from zero.

A good way to compensate for this potential unbalance is to compute the initial bias for each hidden neuron such that its net input is zero for an "average" training case. Let $\bar{x}_j$ be the mean of variable j in the training set. Then the initial bias for hidden neuron i is given by Equation 3-18.

$$c_i = -\sum_j \bar{x}_j W_{ij} \tag{3-18}$$

If the user wants to impose a sparsity penalty (page 101) to encourage specialization among the hidden neurons, then a modification of this equation might be appropriate. Suppose $f(.)$ is the neuron's logistic activation function (Equation 1-4). We leave it as a simple exercise for the reader to confirm that $f(\log(q/(1-q)))=q$. In other words, a net input of $\log(q/(1-q))$ produces an activation probability of q. If q is the user's desired activation probability, then the initial bias could be computed with Equation 3-19. However, it has been my experience that initializing with Equation 3-18 tends to be slightly more stable, and subsequent training does an excellent job of producing the desired sparsity.

$$c_i = \log\left(\frac{q}{1-q}\right) - \sum_j \bar{x}_j W_{ij} \qquad (3\text{-}19)$$

Visible Neuron Bias

The goal in finding a good parameter set for subsequent refinement is to minimize the reconstruction error. Even though this is not the ultimate training goal, the reconstruction error is so highly correlated with the statistical likelihood of a parameter set that beginning maximum likelihood training at a low point of reconstruction error is a virtuous goal.

A necessary (though naturally not sufficient) condition for small reconstruction error is that the probabilities for the reconstructed visible neurons be close to the probabilities for the inputs as observed in the training set. As in the prior section, let $\bar{x}_j$ be the mean of variable j in the user's training set. Recall from the prior section that a visible neuron's net input of $\log(p/(1-p))$ produces an activation probability of p.

In the case of setting the initial bias for hidden neurons, we knew from the training set the probability that each input would be activated. But now we have no such neuron-by-neuron knowledge. Well, that's not quite true. For any trial weight set and accompanying hidden bias vector, we could go through the training set and thereby find the average activation of each hidden neuron. But that would be an awful lot of computation.

A much faster approach is to just let q be the average hidden neuron activation that we expect. This would be 0.5 if Equation 3-18 was used for the hidden biases and would be the specified q if Equation 3-19 was used. Then Equation 3-20 provides reasonable values for the input neuron biases.

$$b_j = \log\left(\frac{\bar{x}_j}{1 - \bar{x}_j}\right) - q \sum_i W_{ij} \qquad (3\text{-}20)$$

Code for Reconstruction Error

Reconstruction error was discussed on page 93. There, we saw that by using Equations 3-3 and 3-4, we could find the hidden neuron probabilities corresponding to a training case and then bounce back to the visible layer. Either Equation 3-5 or 3-6 could be used to compare the original case with its reconstruction and thereby define an error measurement. This section will show code for implementing this algorithm, broken into two separately discussed sections.

The calling parameter list is straightforward, with one small exception. The data array has a row for each of the nc cases, and it has max_neurons columns, but only the first n_inputs of these columns contain data. The reason for this will become clear in the next chapter when I discuss greedy multiple-layer training. The reconstruction error is cumulated across cases, and the first step for each case is to get its data into visible1.

```
double rbm1_threaded (
  int nc ,                  // Number of cases
  int n_inputs ,            // Number of inputs
  int max_neurons ,         // Maximum number of neurons in any layer, as well as nin
  double *data ,            // Nc rows by max_neurons columns of input data; 0-1
  int nhid ,                // Number of hidden neurons
  double *w ,               // Weight matrix, nhid rows, n_inputs cols
  double *in_bias ,         // Input bias vector
  double *hid_bias ,        // Hidden bias vector
  double *visible1 ,        // Work vector n_inputs long
  double *hidden1           // Work vector nhid long
  )
{
  int icase, ihid, ivis ;
  double error, sum, *wptr, *dptr, P ;

  error = 0.0 ; // Will cumulate reconstruction error

  for (icase=0 ; icase<nc ; icase++) {      // Pass through all cases, cumulating error
    dptr = data + icase * max_neurons ;     // Point to this case in the data
    for (ivis=0 ; ivis<n_inputs ; ivis++)
      visible1[ivis] = dptr[ivis] ;
```

The following loop is a straightforward implementation of Equation 3-3. Initialize the sum to the hidden bias and then sum the dot product of the input vector with this hidden neuron's weight vector. Apply the logistic activation function to the neuron's net input.

```
for (ihid=0 ; ihid<nhid ; ihid++) {

  wptr = w + ihid * n_inputs ;      // Weight vector for this neuron
  sum = hid_bias[ihid] ;
  for (ivis=0 ; ivis<n_inputs ; ivis++)
    sum += wptr[ivis] * visible1[ivis] ;
  hidden1[ihid] = 1.0 / (1.0 + exp(-sum)) ;
  }
```

The next loop similarly evaluates Equation 3-4 to get the reconstructed value P for each input. A runtime choice could be made for whether to use the mean squared error of Equation 3-5 or the cross entropy of Equation 3-6. Here I make it a compile-time choice simply because I believe MSE is better and I have no problem with carving it in stone. But by coding both, I can still experiment with both versions.

```
for (ivis=0 ; ivis<n_inputs ; ivis++) {
    sum = in_bias[ivis] ;
    for (ihid=0 ; ihid<nhid ; ihid++)
      sum += w[ihid*n_inputs+ivis] * hidden1[ihid] ;
    P = 1.0 / (1.0 + exp(-sum)) ;
#if RECON_ERR_XENT
    error -= visible1[ivis] * log(P+1.e-10) + (1.0 - visible1[ivis]) * log(1.0-P+1.e-10) ;
#else
    double diff ;
    diff = visible1[ivis] - P ;
    error += diff * diff ;
#endif
    }

  } // For icase

  return error ;
}
```

Multithreading Initial Weight Selection

The code presented in this section covers both generating the trial parameter sets and evaluating their reconstruction errors with a multithreaded algorithm. Many concepts vital to multithreading in Windows were covered starting on page 29. Readers should review that material because many of the issues presented there will be glossed over in this section to avoid redundancy.

The parameters, addresses of work areas, and so forth, are passed to the threaded routine via the following structure. The wrapper routine follows.

```
typedef struct {
   int nc ;                      // Number of cases
   int n_inputs ;                // Number of inputs
   int max_neurons ;             // Maximum number of neurons in any layer, including input
   double *data ;                // Nc rows by max_neurons columns of input data; 0-1
   int nhid ;                    // Number of hidden neurons
   double *w ;                   // Weight matrix; nhid sets of n_inputs weights
   double *in_bias ;             // Input bias vector
   double *hid_bias ;            // Hidden bias vector
   double *visible1 ;            // Work vector n_inputs long
   double *hidden1 ;             // Work vector nhid long
   double crit ;                 // Computed criterion returned here
} RBM_THR1_PARAMS ;

static unsigned int __stdcall rbm1_wrapper ( LPVOID dp )
{
  ((RBM_THR1_PARAMS *) dp)->crit = rbm1_threaded (
               ((RBM_THR1_PARAMS *) dp)->nc ,
               ((RBM_THR1_PARAMS *) dp)->n_inputs ,
               ((RBM_THR1_PARAMS *) dp)->max_neurons ,
               ((RBM_THR1_PARAMS *) dp)->data ,
               ((RBM_THR1_PARAMS *) dp)->nhid ,
               ((RBM_THR1_PARAMS *) dp)->w ,
               ((RBM_THR1_PARAMS *) dp)->in_bias ,
               ((RBM_THR1_PARAMS *) dp)->hid_bias ,
```

```
                    ((RBM_THR1_PARAMS *) dp)->visible1 ,
                    ((RBM_THR1_PARAMS *) dp)->hidden1 ) ;
  return 0 ;
}
```

The calling parameter list is similar to that for the reconstruction error routine already presented. However, there is one vital difference. The prior routine, which is called for each thread, must have its own private copy of parameters (weights and biases) as well as work areas (visible1 and hidden1). Otherwise, different threads would overwrite one another's data and create mass chaos. Thus, when this routine is called, these vectors must be allocated with sufficient space for max_threads sets of parameters and work areas. Here is the calling parameter list and local variable declarations:

```
double rbm_thr1 (
  int nc ,                   // Number of cases
  int n_inputs ,             // Number of inputs
  int max_neurons ,          // Maximum number of neurons in any layer, including input
  double *data ,             // Nc rows by max_neurons columns of input data; 0-1
  int nhid ,                 // Number of hidden neurons
  double *w ,                // Returned weight matrix, nhid sets of n_inputs weights;
                             // max_threads sets of them
  double *in_bias ,          // Returned input bias vector; max_threads sets
  double *hid_bias ,         // Returned hidden bias vector; max_threads sets
  double *visible1 ,         // Work vector n_inputs long; max_threads sets
  double *hidden1 ,          // Work vector nhid long; max_threads sets
  double *in_bias_best ,     // Work vector n_inputs long
  double *hid_bias_best ,    // Work vector nhid long
  double *w_best ,           // Work vector n_inputs * nhid long
  double *data_mean          // Work vector n_inputs long
  )
{
  int irand, ivis, ihid, i, k, n_rand, n_threads, empty_slot, ret_val ;
  double error, best_err, sum, wt, *dptr, *wptr, *hid_bias_ptr, *in_bias_ptr, diff ;
  char msg[4096] ;
  RBM_THR1_PARAMS params[MAX_THREADS] ;
  HANDLE threads[MAX_THREADS] ;
```

The mean of each input variable is needed for computing reasonable bias vectors given a random weight set. But because we will be taking the log of these means and one minus the means, they must be kept away from zero and one. This is done in the next section of code, shown here:

```
for (ivis=0 ; ivis<n_inputs ; ivis++)
  data_mean[ivis] = 0.0 ;

for (i=0 ; i<nc ; i++) {                    // Pass through all cases, cumulating mean vector
  dptr = data + i * max_neurons ;    // Point to this case in the data
  for (ivis=0 ; ivis<n_inputs ; ivis++)
    data_mean[ivis] += dptr[ivis] ;
  }

for (ivis=0 ; ivis<n_inputs ; ivis++) {
  data_mean[ivis] /= nc ;
  if (data_mean[ivis] < 1.e-8)
    data_mean[ivis] = 1.e-8 ;
  if (data_mean[ivis] > 1.0 - 1.e-8)
    data_mean[ivis] = 1.0 - 1.e-8 ;
  }
```

Most of the parameters needed by the routine that computes the reproduction error remain constant for each thread invocation (trial of a weight set). Some of them, such as the number of cases, inputs, and hidden neurons, are the same for every thread. Others, such as addresses of work areas, are different for each thread to avoid trespassing on one another's private property. These assignments are done now.

```
n_rand = TrainParams.n_rand ;   // Number of random weight sets to test

for (i=0 ; i<max_threads ; i++) {
  params[i].nc = nc ;
  params[i].n_inputs = n_inputs ;
  params[i].max_neurons = max_neurons ;
  params[i].nhid = nhid ;
  params[i].data = data ;
  params[i].visible1 = visible1 + i * max_neurons ;
  params[i].hidden1 = hidden1 + i * max_neurons ;
  params[i].w = w + i * nhid * n_inputs ;
```

```
    params[i].hid_bias = hid_bias + i * max_neurons ;
    params[i].in_bias = in_bias + i * max_neurons ;
    }
```

We do some initializations and begin the "endless" loop that will start separate threads for weight set trials. The global variable max_threads is the maximum number of threads that will be allowed. Larger values will require more memory, as noted in the calling parameter list. We will do n_rand trials, and irand will count them. The software thread to be started is indexed by k. While the initial queue is filling, k just counts up. But after the queue is filled and threads are starting and completing, k is set to the thread that just finished its work and returned, ready for a new task.

```
  n_threads = 0 ;          // Counts threads that are active
  for (i=0 ; i<max_threads ; i++)
    threads[i] = NULL ;

  irand = 0 ;              // Index of try
  empty_slot = -1 ;        // After full, will identify the thread that just completed
  best_err = 1.e40 ;       // Will keep track of best reconstruction error

  for (;;) {               // Main thread loop processes all tries

    if (irand < n_rand) {  // If there are still some weight trials to do
      if (empty_slot < 0)  // Negative while we are initially filling the queue
        k = n_threads ;
      else
        k = empty_slot ;   // This thread has finished and can now be used
```

At this point we are ready to generate a trial weight set and bias vectors for the hidden and visible neurons. This trial run will be performed by thread k, so we must get the pointers to the parameter areas for this thread. Then we use Equation 3-16 to compute diff as the "grand variation" for this trial. By using a separate grand variation for each trial, we avoid the problem of a homogeneous search, with all trials having small weights or all trials having at least a few large weights.

```
    wptr = params[k].w ;
    hid_bias_ptr = params[k].hid_bias ;
    in_bias_ptr = params[k].in_bias ;

    diff = 4.0 * unifrand_fast() / sqrt ( sqrt ( (double) n_inputs * nhid ) ) ;
```

Now we can use Equation 3-17 to get trial weights and then use Equation 3-18 to compute the associated hidden neuron bias. I chose to forego Equation 3-19 simply because my (limited) experience is that the randomness inherent in the trial generation process can occasionally put neurons out in left field if Equation 3-19 is used to initially unbalance activation. However, users should feel free to experiment on their own. I could easily be wrong on this choice.

```
for (ihid=0 ; ihid<nhid ; ihid++) {
   sum = 0.0 ;

   for (ivis=0 ; ivis<n_inputs ; ivis++) {    // Get visible weights for this hidden neuron
      wt = diff * (unifrand_fast() - 0.5) ;    // Equation 3-17
      wptr[ihid*n_inputs+ivis] = wt ;
      sum += data_mean[ivis] * wt ;            // We'll need this for Equation 3-18
      }

   hid_bias_ptr[ihid] = -sum ;                 // Equation 3-18
   } // For ihid
```

The input neuron bias vector is computed with Equation 3-20.

```
for (ivis=0 ; ivis<n_inputs ; ivis++) {     // Also center the visible
   sum = 0.0 ;
   for (ihid=0 ; ihid<nhid ; ihid++)
      sum += wptr[ihid*n_inputs+ivis] ;
   in_bias_ptr[ivis] = log ( data_mean[ivis] / (1.0 - data_mean[ivis]) ) - 0.5 * sum ;
   }
```

We can now start the thread that tests this trial parameter set. Increment the counter of the number of threads executing, and also increment the counter of the number of trials tested.

```
threads[k] = (HANDLE) _beginthreadex ( NULL , 0 , rbm1_wrapper , &params[k] ,
                              0 , NULL ) ;
++n_threads ;
++irand ;
} // if (irand < n_rand)
```

At this point, one of three situations exists.

- We are just beginning, still filling the queue of threads.

- The maximum allowable number of threads are running, and we still have more trial parameter sets to test.

- All trial parameter sets have had threads started, and now we are just waiting for the threads to finish.

The first case does not require anything to be done. We just let the "endless" loop iterate again to generate a new trial parameter set and start its thread. The next block of code handles the second case. The timeout constant of 12000000 milliseconds must be large enough to give the threads time to finish but small enough that the user is not left clueless when the computer apparently freezes. Finally, we set empty_slot to the slot just freed for reuse, close the thread, and decrement the counter of threads executing.

```
if (n_threads == max_threads && irand < n_rand) {
  ret_val = WaitForMultipleObjects ( n_threads , threads , FALSE , 12000000 ) ;
  error = params[ret_val].crit ;

  // If we just improved, save the best-so-far parameters
  if (error < best_err) {
    best_err = error ;
    for (ihid=0 ; ihid<nhid ; ihid++) {
      hid_bias_best[ihid] = params[ret_val].hid_bias[ihid] ;
      for (ivis=0 ; ivis<n_inputs ; ivis++)
        w_best[ihid*n_inputs+ivis] = params[ret_val].w[ihid*n_inputs+ivis] ;
      }

    for (ivis=0 ; ivis<n_inputs ; ivis++)
      in_bias_best[ivis] = params[ret_val].in_bias[ivis] ;
    }

  empty_slot = ret_val ;
  CloseHandle ( threads[empty_slot] ) ;
  threads[empty_slot] = NULL ; // Need this for user escape code not shown here
  --n_threads ;
  }
```

The third case is similar to the second but a little easier. We wait until all of the threads are finished. Then we must check the criterion of each of the n_threads threads that we just waited for and keep track of the best. Finally, we close all of these threads and break out of the "endless" loop that we have been in.

```
else if (irand == n_rand) {
  ret_val = WaitForMultipleObjects ( n_threads , threads , TRUE , 1200000 ) ;

  for (i=0 ; i<n_threads ; i++) {
    error = params[i].crit ;

    // If we just improved, save the best-so-far parameters
    if (error < best_err) {
      for (ihid=0 ; ihid<nhid ; ihid++) {
        hid_bias_best[ihid] = params[i].hid_bias[ihid] ;
        best_err = error ;
        for (ivis=0 ; ivis<n_inputs ; ivis++)
          w_best[ihid*n_inputs+ivis] = params[i].w[ihid*n_inputs+ivis] ;
        }
      for (ivis=0 ; ivis<n_inputs ; ivis++)
        in_bias_best[ivis] = params[i].in_bias[ivis] ;
      }
    CloseHandle ( threads[i] ) ;
    } // For i, processing all threads just returned
  break ; // All done so break out of endless loop
  } // Waiting for final threads to finish
```

The final step is to copy the best parameters for the user and return the reconstruction error, scaled per the number of cases and inputs.

```
for (ihid=0 ; ihid<nhid ; ihid++) {
  hid_bias[ihid] = hid_bias_best[ihid] ;
  for (ivis=0 ; ivis<n_inputs ; ivis++)
    w[ihid*n_inputs+ivis] = w_best[ihid*n_inputs+ivis] ;
  }
for (ivis=0 ; ivis<n_inputs ; ivis++)
  in_bias[ivis] = in_bias_best[ivis] ;

return best_err / (nc * n_inputs) ;
```

Stochastic Gradient Descent Basic Principles

This section discusses some general principles of training an RBM that I have found to be useful. Please understand that this field is still young and evolving. Other researchers likely have even better ideas than those presented here, and still better refinements are being developed regularly. But the algorithms given here are well established and stable and are a solid foundation for more exotic approaches.

The Core Algorithm

The idea behind stochastic gradient descent is that you compute an estimate of the gradient for some parameter set and then adjust the parameters by moving slightly in the direction opposite the gradient. At least for small moves, the error criterion will, on average, decrease. This simple algorithm is shown in Equations 3-21 and 3-22. In these equations, W_k is the weight set (including bias vectors) as of update k, *rate* is a very small number called the *learning rate*, and Δ_k is the gradient at W_k.

$$increment_k = -rate * \Delta_k \qquad (3\text{-}21)$$

$$W_{k+1} = W_k + increment_k \qquad (3\text{-}22)$$

One problem with this simple approach is that when the error criterion function is a long, narrow chasm, which happens often, the updates will bounce back and forth across the walls of the chasm, making little net progress with each bounce. For this reason, a *momentum* term is usually included to propel adjustments in the average direction of movement. Each increment includes a fraction of the increment from the prior update, as shown in Equation 3-23. We can see the effect of momentum in Figure 3-2. Observe that without momentum the path would be perpendicular to the level curve. But by including an extra shove in the direction of the prior update, we push the net movement deeper into the chasm.

$$increment_k = mom * increment_{k-1} - rate * \Delta_k \qquad (3\text{-}23)$$

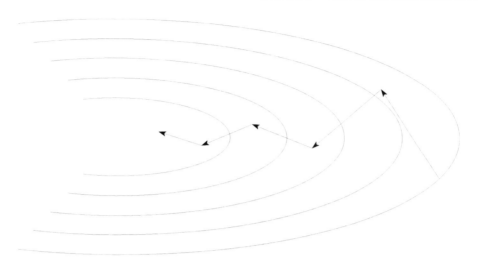

Figure 3-2. *Momentum pushes rapidly into the chasm*

Dividing Epochs into Batches

I discussed this subject in the context of supervised training on page 82. However, it is particularly relevant for RBM training, so we'll repeat the discussion here. One could evaluate the gradient for each case, one at a time, and apply a small update to the weights accordingly each time. However, this would incur enormous overhead.

At the other extreme, we could sum the gradients for all cases and apply a single weight update after passing through the entire training set. This would have minimal overhead, and moreover, the error variance of the gradient would be relatively small because random errors would largely cancel when we sum the individual gradient estimates. Having an accurate gradient estimate is great.

Between these two extremes we could divide the training set into batches. Sum the gradient for each batch and do one update per batch. But how large should we make the batches? An important consideration is that the trade-off between gradient evaluation time per batch and number of batches required for satisfactory training is not a net wash. In particular, when we decrease the batch size and hence decrease the compute time per batch, the number of batches required for convergence does not increase as fast as the gradient time decreases. Pulling some numbers out of the air, we may find that if we decrease the batch size (and hence compute time per batch) by a factor of ten, we may find that five times as many batches are required because of the increased error of the gradient in each batch. But still, we are ahead by a factor of two.

This, of course, is offset by the increased overhead of using numerous batches. The overhead for starting a thread under Windows is quite small, so it is not much of a consideration. On the other hand, the overhead of launching a CUDA kernel is significant, so we might be inclined to use larger batches with CUDA processing than with Windows multithreading.

Shuffling Epochs

If we were to process the entire training set in one huge batch, the order in which training cases appear in the dataset would be irrelevant. However, because in most practical situations we will be separating the dataset into batches as described in the prior section, the order of training cases will make a difference.

There are two order-dependent issues to consider. First, suppose the training set has significant serial correlation. This may happen, for example, if the data involves time series. It may also happen if the experimenter has deliberately varied some aspect of the test, with cases derived from different test conditions grouped together. If there is serial correlation, we may find that with one batch the gradient moves in one direction, and with the next batch it moves back the other way. This is not good.

Second, even without overt serial dependency, the training set may contain one or more very unusual cases or accidental groupings of like cases. It is better not only to disperse unusual cases or clusters of patterns throughout batches, but even to vary this dispersion from epoch (pass through the entire training set) to epoch so that we don't keep getting the same gradient backtracking each time.

Of course, in practice it makes little sense to shuffle the actual cases; this would be too expensive if there are more than a few variables. Rather, a programmer should address cases through a vector of indices into the dataset and then just shuffle the index vector.

Updating the Learning Rate and Momentum

The simplest way to handle learning rate and momentum is to fix them at preordained values throughout training. The problem with this approach is that if they are too small, convergence to a minimum criterion will be virtually assured but will also take far longer than need be. Thus, one is inclined to make them large. But if the learning rate is too large, the minimum will be overshot repeatedly, and convergence will be difficult if not impossible. And if the momentum is too large, the path through parameter space will not be able to turn corners well and will shoot right through sharp curves (which *does* happen).

For these reasons, more advanced methods were devised for determining learning rate and momentum. One might set the learning rate to change at a steady rate as training progresses or set it to be a constant fraction of the average weight size. Similarly, one might assume that as training progresses, the weights will be more and more contained in a well-behaved canyon and thus increase momentum steadily.

But these are disturbingly heuristic approaches. The truly optimal learning rate is a function of the Hessian matrix, which is impractical to compute. And simpler methods such as directional minimization and conjugate gradients (page 84), while excellent when deterministic gradients are available, are worthless when the gradient estimates include substantial random variation, as is the case here. What to do?

In my own work I have found a simple yet highly effective method for achieving nearly optimal learning rates and at least reasonable momentum factors. The motivation for this technique is the idea that if the learning rate is too small, successive gradient estimates will point in nearly the same direction. Conversely, if the rate is too high, we will overshoot the minimum in this direction, and hence the gradient will point us right back toward where we came from. Thus, we use the cosine of the angle between the current gradient and the prior gradient as an indicator of the quality of the learning rate. If the cosine is very small (the gradients are pointing in nearly the same direction), the learning rate should be increased, and if it is very large (the gradients are pointing in nearly opposite directions), we need to cut the learning rate.

The cosine of the angle between two vectors is their dot product divided by the product of their lengths, as shown in Equation 3-24.

$$GradCos_k = \frac{\Delta_k \cdot \Delta_{k-1}}{\|\Delta_k\|\|\Delta_{k-1}\|} \tag{3-24}$$

This same measure can be used to keep tabs on the momentum factor, although not with as much efficacy as for the learning rate. The idea is that when things are going well, the successive gradients will be somewhat perpendicular. They need not be exactly perpendicular, as is illustrated in Figure 2-6. But if the learning rate is reasonable yet the angle is excessively large or small, the momentum is the likely suspect. Remember that momentum is really a departure from the direction of steepest descent and hence is a dangerous source of instability. You are always safe using little or no momentum; the only price paid is slower convergence. But excessive momentum can prevent convergence or lead to a local minimum far out in left field. In case of doubt, cut momentum.

This leads to a simple algorithm for determining momentum: start out small and slowly but steadily increase it as training progresses. But if the absolute value of Equation 3-24 becomes excessive, slash the momentum. And avoid the temptation to increase the momentum if the angle is small; this leads to instability. Instead, increase the learning rate and then let the momentum naturally increase. This seems to work well in practice.

Determining Convergence

When one is doing deterministic minimization, determining convergence is straightforward. One can look at the rate at which the criterion is decreasing and stop when the rate drops below a specified threshold. Or one can look at the length of the gradient vector and stop when the length becomes sufficiently small.

These approaches are problematic for training an RBM. Except for tiny toy applications, we cannot compute the likelihod criterion that we want to optimize, so we cannot use its rate of decrease as a stopping test. It's tempting to use the reconstruction error (page 93) for the test and stop iterating when its rate of decrease drops to nearly zero. This is not totally inappropriate because this value has solid intuitive appeal and is often highly correlated with the likelihood. But this is not the function whose gradient we are computing, so gradient descent does not necessarily correspond to a reduction of the reconstruction error.

This is especially true if we perform many iterations of contrastive divergence (page 97), which produces a better approximation to the likelihood gradient. If we do just one iteration, the correspondence between the gradient and the reconstruction error is usually close, but even then it's far from perfect. And as training progresses, the correspondence becomes progressively worse. This is because the weights generally increase in magnitude as we train longer, and larger weights result in slower mixing rate, meaning that the reconstruction error can be small while the model is still far from providing a good representation of the distribution of the training data. So, if we stop training when the rate of decrease of the reconstruction error drops to near zero, we are probably stopping much too soon.

The other reason that stopping criteria for deterministic problems are inappropriate for RBM training is that the gradient is stochastic; even if the weights happen to be at exactly the optimum, the gradient will still jump around because of the impact of randomness in setting hidden (and perhaps visible) neuron activations. Conversely, a gradient of essentially zero need not correspond to an optimum weight set.

Still, it's possible to use a variation of this measure as a fairly good stopping test. By looking at the gradient element having maximum magnitude, we can have a measure of the worst that randomness can do when we are near the optimum. Then we compare this to the size of the largest magnitude weight (not necessarily the weight corresponding to the largest gradient element) because we want a measure of convergence relative to how much any input can impact hidden neuron activations.

Another benefit of this test is that if one or more weights are blowing up (a fairly unusual pathological situation, especially if a weight penalty is in force), we probably want to stop.

That test should be secondary, though. I have found that the best measure of convergence is the number of times *in a row* for which this gradient-to-weight ratio fails to decrease from its minimum so far. This counting test can fail in the rare instance that some weight is blowing up, which is why the ratio test just described should be included. But under normal conditions, once we are near a likelihood maximum, randomness will at some point provide a small value for the gradient/weight ratio, which is difficult to further reduce in subsequent epochs. So if we find that a considerable number of epochs have passed with no reduction of this ratio, we can conclude that we are close to an optimal weight configuration. This should be the primary convergence criterion because it is quite effective.

Code for Multithreaded RBM Training

We now present code (minus error checking and so on) for approximate maximum-likelihood training of an RBM using contrastive divergence. It will be shown in sections, alternating with explanations. The complete source code for this module is available for free download from the author's web site. The calling parameter list for the gradient routine (algorithm is on page 99) is given first. The constants at the beginning are for fast random number generation later.

```
#define IA 16807
#define IM 2147483647
#define AM (1.0 / IM)
#define IQ 127773
#define IR 2836
```

```
static void rbm2_threaded (
    int istart ,              // First case in this batch
    int istop ,               // One past last case
    int ncols ,               // Number of columns in data
    int n_inputs ,            // Number of inputs
    double *data ,            // 'Training cases' rows by ncols columns of input data; 0-1
    int nhid ,                // Number of hidden neurons
    int n_chain ,             // Length of Markov chain
    int mean_field ,          // Use mean field instead of random sampling?
    double *w ,               // Weight matrix, nhid sets of n_inputs weights
    double *in_bias ,         // Input bias vector
    double *hid_bias ,        // Hidden bias vector
    int *shuffle_index ,      // For addressing shuffled data
    double *visible1 ,        // Work vector n_inputs long
    double *visible2 ,        // Work vector n_inputs long
    double *hidden1 ,         // Work vector nhid long
    double *hidden2 ,         // Work vector nhid long
    double *hidden_act ,      // Work vector nhid long
    double *in_bias_grad ,    // Cumulates gradient here
    double *hid_bias_grad ,   // Cumulates gradient here
    double *w_grad ,          // Cumulates gradient here
    double *hid_on_frac ,     // Cumulates fraction of time each hidden neuron is on
    double *error             // Cumulates reconstruction criterion
    )
```

Most of these parameters should be self-explanatory, especially for readers who have studied prior modules. As with the routine for finding starting weights, the training data array has ncols columns, with the first n_inputs of them being used. The reason for this will become clear in the next chapter when I present greedy learning.

The array shuffle_index is as long as there are cases in the complete training set, and it contains the integers zero through *number of cases* minus one. These will be shuffled by the calling routine and used in this routine to access the cases in random order.

Next are the variable declarations and initialization for the random number generator. By including istop and shuffle_index[0] in the random initialization, we ensure that for each batch and epoch we get a different sequence of random numbers for determining hidden neuron activations. One peculiarity of this random number generator is that the value zero must never be used for initialization, so we explicitly prevent this.

```
int k, randnum, icase, ivis, ihid, ichain ;
double sum, *wptr, *dptr, P, Q, frand ;

randnum = (istop + shuffle_index[0]) % IM ;
if (randnum == 0)
  randnum = 1 ;
```

We zero the areas in which the gradient and error will be cumulated for this batch. The array hid_on_frac tallies the fraction of time each hidden neuron is on. This is needed later for sparsity penalties.

```
for (ihid=0 ; ihid<nhid ; ihid++) {
  hid_bias_grad[ihid] = 0.0 ;
  hid_on_frac[ihid] = 0.0 ;
  for (ivis=0 ; ivis<n_inputs ; ivis++)
    w_grad[ihid*n_inputs+ivis] = 0.0 ;
  }

for (ivis=0 ; ivis<n_inputs ; ivis++)
  in_bias_grad[ivis] = 0.0 ;

*error = 0.0 ;
```

The main loop processes each case in this batch, using the algorithm listed on page 99 to estimate the gradient for each case and summing the gradient across the batch. The first step in this algorithm is to use Equation 3-3 to compute the activation probabilities of the hidden neurons. We'll save these in hidden1 for use in the positive gradient term and also put them into hidden2 for sampling in the Monte Carlo chain.

```
for (icase=istart ; icase<istop ; icase++) {
  dptr = data + shuffle_index[icase] * ncols ;      // Point to this case in the data
  for (ivis=0 ; ivis<n_inputs ; ivis++)             // Get this case
    visible1[ivis] = dptr[ivis] ;
```

```
for (ihid=0 ; ihid<nhid ; ihid++) {
  wptr = w + ihid * n_inputs ;                  // Weight vector for this neuron
  sum = hid_bias[ihid] ;
  for (ivis=0 ; ivis<n_inputs ; ivis++)         // Equation 3-3 summation
    sum += wptr[ivis] * visible1[ivis] ;
  Q = 1.0 / (1.0 + exp(-sum)) ;                 // Probability = f(sum)
  hidden1[ihid] = hidden2[ihid] = Q ;           // We'll need hidden2 for CD-k loop below
  hid_on_frac[ihid] += Q ;                      // Need this for sparsity penalty
  }
```

We can optionally compute the reconstruction error here using the slow, accurate method. We can also use either mean squared error (Equation 3-5) or cross entropy (Equation 3-6). These could be user options.

```
#if RECON_ERR_DIRECT
  // Compute the reconstruction error the deterministic but expensive way
  for (ivis=0 ; ivis<n_inputs ; ivis++) {       // This loop is not needed for gradient
    sum = in_bias[ivis] ;                       // It is only for reconstruction error
    for (ihid=0 ; ihid<nhid ; ihid++)           // and may be an expensive waste
      sum += w[ihid*n_inputs+ivis] * hidden1[ihid] ; // because it is a nested loop!
    P = 1.0 / (1.0 + exp(-sum)) ;               // Equation 3-4
#if RECON_ERR_XENT
    *error -= visible1[ivis] * log(P+1.e-10) + (1.0 - visible1[ivis]) * log(1.0-P+1.e-10) ;
#else
    double diff = visible1[ivis] - P ;
    *error += diff * diff ;
#endif
    }
#endif
```

We now run the Markov chain that gives us successively better approximations to a random sample from the "natural" RBM distribution corresponding to the current weight set. Each time the loop begins, hidden2 contains the probabilities of hidden neuron activations. We generate a random number and sample to get hidden_act.

```
for (ichain=0 ; ichain<n_chain ; ichain++) {
  for (ihid=0 ; ihid<nhid ; ihid++) {
    k = randnum / IQ ;
    randnum = IA * (randnum - k * IQ) - IR * k ;
    if (randnum < 0)
      randnum += IM ;
    frand = AM * randnum ;
    hidden_act[ihid] = (frand < hidden2[ihid]) ? 1.0 : 0.0 ;
  }
```

Compute the visible activation probabilities using Equation 3-4. And now that these probabilities are available, we can compute the reconstruction error as either mean squared error (Equation 3-5) or cross entropy (Equation 3-6). The probabilities P are based on sampled hidden activations, so this measure is not as accurate as the deterministic version, but we need these probabilities for the gradient anyway, so there is no wasted effort as there would be for the deterministic method.

```
for (ivis=0 ; ivis<n_inputs ; ivis++) {
  sum = in_bias[ivis] ;
  for (ihid=0 ; ihid<nhid ; ihid++)
    sum += w[ihid*n_inputs+ivis] * hidden_act[ihid] ;
  P = 1.0 / (1.0 + exp(-sum)) ; // Equation 3-4

#if ! RECON_ERR_DIRECT
  if (ichain == 0) {
#if RECON_ERR_XENT
    *error -= visible1[ivis] * log(P+1.e-10) + (1.0-visible1[ivis]) * log(1.0-P+1.e-10) ;
#else
    double diff = visible1[ivis] - P ;
    *error += diff * diff ;
#endif
  }
#endif
```

The last step in the "visible neuron" loop is to either just save the probability directly or, if not mean_field, generate a random number and use it to sample the reconstructed visible activation.

```
if (mean_field)
  visible2[ivis] = P ;
else {
  k = randnum / IQ ;
  randnum = IA * (randnum - k * IQ) - IR * k ;
  if (randnum < 0)
    randnum += IM ;
  frand = AM * randnum ;
  visible2[ivis] = (frand < P) ? 1.0 : 0.0 ; // Sample the activation
  }
} // For each visible neuron
```

The last step in the Markov chain is to use the visible neuron values, either the probabilities or the sampled activations, to compute the hidden neuron activation probabilities.

```
for (ihid=0 ; ihid<nhid ; ihid++) {
  wptr = w + ihid * n_inputs ;     // Weight vector for this neuron
  sum = hid_bias[ihid] ;
  for (ivis=0 ; ivis<n_inputs ; ivis++)
    sum += wptr[ivis] * visible2[ivis] ;
  hidden2[ihid] = 1.0 / (1.0 + exp(-sum)) ;
  }
} // For Markov chain
```

At this point we have the "data" and "model" visible and hidden neuron probabilities and/or activations. We can then use Equation 3-12 to compute the (negative) gradient. This equation is in terms of expectations. The "data" values we now have are unbiased random samples from the training set distribution. The "model" values are biased, with the degree of bias decreasing as the Monte Carlo chain repeats. But they are nevertheless reasonable. An entire batch of samples, when averaged, will be an acceptable approximation to the expectations in Equation 3-12. Note that if we are not using a mean field approximation, we must sample the "model" hidden activations.

```
for (ihid=0 ; ihid<nhid ; ihid++) {

   if (mean_field) {
     hid_bias_grad[ihid] += hidden1[ihid] - hidden2[ihid] ;
     for (ivis=0 ; ivis<n_inputs ; ivis++)
        w_grad[ihid*n_inputs+ivis] += hidden1[ihid] * visible1[ivis] -
                                      hidden2[ihid] * v isible2[ivis] ;

     }

   else {
     k = randnum / IQ ;
     randnum = IA * (randnum - k * IQ) - IR * k ;
     if (randnum < 0)
        randnum += IM ;
     frand = AM * randnum ;
     hidden_act[ihid] = (frand < hidden1[ihid]) ? 1.0 : 0.0 ;
     hid_bias_grad[ihid] += hidden_act[ihid] - hidden2[ihid] ;
     for (ivis=0 ; ivis<n_inputs ; ivis++)
        w_grad[ihid*n_inputs+ivis] += hidden_act[ihid] * visible1[ivis] -
                                      hidden2[ihid] * v isible2[ivis] ;

     }
   }

   for (ivis=0 ; ivis<n_inputs ; ivis++)
     in_bias_grad[ivis] += visible1[ivis] - visible2[ivis] ;

   } // For each case in this batch
}
```

The routine just shown is the gradient computation workhorse that will be called in threads. Threading has been discussed several times already, so we will rush through it this time, trusting that the reader will review the prior discussions if needed to understand what's going on. We begin with the data structure used to pass parameters and the wrapper that is called for each thread. These are listed here:

```
typedef struct {
   int istart ;              // First case in this batch
   int istop ;               // One past last case
   int ncols ;               // Number of columns in data
```

```
   int n_inputs ;          // Number of inputs
   double *data ;          // 'Training cases' rows by ncols columns of input data; 0-1
   int nhid ;              // Number of hidden neurons
   int n_chain ;           // Length of Markov chain; often 1
   int mean_field ;        // Use mean field instead of random sampling?
   double *w ;             // Weight matrix; nhid sets of n_inputs weights
   double *in_bias ;       // Input bias vector
   double *hid_bias ;      // Hidden bias vector
   int *shuffle_index ;    // For addressing shuffled data
   double *visible1 ;      // Work vector n_inputs long
   double *visible2 ;      // Work vector n_inputs long
   double *hidden1 ;       // Work vector nhid long
   double *hidden2 ;       // Work vector nhid long
   double *hidden_act ;    // Work vector nhid long
   double *in_bias_grad ;  // Cumulates gradient here
   double *hid_bias_grad ; // Cumulates gradient here
   double *w_grad ;        // Cumulates gradient here
   double *hid_on_frac ;   // Cumulates fraction of time each hidden neuron is on
   double *error ;         // Cumulates MSE
} RBM_THR2_PARAMS ;

static unsigned int __stdcall rbm2_wrapper ( LPVOID dp )
{
rbm2_threaded (
               ((RBM_THR2_PARAMS *) dp)->istart ,
               ((RBM_THR2_PARAMS *) dp)->istop ,
               ((RBM_THR2_PARAMS *) dp)->ncols ,
               ((RBM_THR2_PARAMS *) dp)->n_inputs ,
               ((RBM_THR2_PARAMS *) dp)->data ,
               ((RBM_THR2_PARAMS *) dp)->nhid ,
               ((RBM_THR2_PARAMS *) dp)->n_chain ,
               ((RBM_THR2_PARAMS *) dp)->mean_field ,
               ((RBM_THR2_PARAMS *) dp)->w ,
               ((RBM_THR2_PARAMS *) dp)->in_bias ,
               ((RBM_THR2_PARAMS *) dp)->hid_bias ,
               ((RBM_THR2_PARAMS *) dp)->shuffle_index ,
```

```
                ((RBM_THR2_PARAMS *) dp)->visible1 ,
                ((RBM_THR2_PARAMS *) dp)->visible2 ,
                ((RBM_THR2_PARAMS *) dp)->hidden1 ,
                ((RBM_THR2_PARAMS *) dp)->hidden2 ,
                ((RBM_THR2_PARAMS *) dp)->hidden_act ,
                ((RBM_THR2_PARAMS *) dp)->in_bias_grad ,
                ((RBM_THR2_PARAMS *) dp)->hid_bias_grad ,
                ((RBM_THR2_PARAMS *) dp)->w_grad ,
                ((RBM_THR2_PARAMS *) dp)->hid_on_frac ,
                ((RBM_THR2_PARAMS *) dp)->error ) ;
  return 0 ;
}
```

The calling parameter list for the training routine is shown next. Most of these parameters have been seen before, so they need no explanation. However, there are a few items that may need explanation.

- As noted on page 100, if one makes just a single pass through the (very expensive!) Markov chain loop, the resulting "sample" from the model distribution will be significantly biased toward the data distribution. Nonetheless, it will contain a large amount of gradient information, more than enough to quickly pull us toward the maximum likelihood weights. Thus, we are inspired to start with a small number of passes, n_chain_start, typically 1, and work toward a somewhat larger number, n_chain_end, at an exponential smoothing rate of n_chain_rate.

- The two convergence criteria, max_no_improvement and convergence_crit, were discussed on page 120.

- The learning_rate parameter is an initial value only. It will be automatically adjusted, as discussed on page 118. The start_momentum and end_momentum parameters are discussed in that section as well.

- Many of the work vectors need to have separate copies for each thread. Therefore, the caller must allocate for the apparent required size times the maximum number of threads possible.

```
double rbm_thr2 (
   int nc ,                      // Number of training cases
   int ncols ,                   // Number of columns in data
   double *data ,                // Nc rows by ncols columns of input data; 0-1
   int n_inputs ,                // Number of inputs
   int nhid ,                    // Number of hidden neurons
   int max_neurons ,             // Maximum number of neurons in any layer
   int n_chain_start ,           // Starting length of Markov chain, generally 1
   int n_chain_end ,             // Ending length of Markov chain
   double n_chain_rate ,         // Exponential smoothing rate for n_chain
   int mean_field ,              // Use mean field instead of random sampling?
   int n_batches ,               // Number of batches per epoch
   int max_epochs ,              // Maximum number of epochs
   int max_no_improvement ,      // Converged if this many epochs no improvement
   double convergence_crit ,     // Convergence criterion for max inc / max weight
   double learning_rate ,        // Learning rate (starting value; quickly adjusted)
   double start_momentum ,       // Learning momentum start value
   double end_momentum ,         // Learning momentum end value
   double weight_penalty ,       // Weight penalty
   double sparsity_penalty ,     // Sparsity penalty
   double sparsity_target ,      // Sparsity target
   double *w ,                   // Computed weight matrix, nhid sets of n_inputs wts
   double *in_bias ,             // Computed input bias vector
   double *hid_bias ,            // Computed hidden bias vector
   int *shuffle_index ,          // Work vector nc long
   double *data_mean ,           // Work vector n_inputs long
   double *visible1 ,            // Work vector n_inputs * max_threads long
   double *visible2 ,            // Work vector n_inputs * max_threads long
   double *hidden1 ,             // Work vector nhid * max_threads long
   double *hidden2 ,             // Work vector nhid * max_threads long
   double *hidden_act ,          // Work vector nhid * max_threads long
   double *hid_on_frac ,         // Work vector nhid * max_threads long
   double *hid_on_smoothed ,     // Work vector nhid long
   double *in_bias_inc ,         // Work vector n_inputs long
   double *hid_bias_inc ,        // Work vector nhid long
   double *w_inc ,               // Work vector n_inputs * nhid long
```

```
double *in_bias_grad ,          // Work vector n_inputs * max_threads long
double *hid_bias_grad ,         // Work vector nhid * max_threads long
double *w_grad ,                // Work vector n_inputs * nhid * max_threads long
double *w_prev                  // Work vector n_inputs * nhid long
)
```

The variable declarations are shown here. The algorithm contains nested subsetting; in other words, each epoch (a complete pass through all the training data) is broken up into batches for weight update, and each batch is distributed across multiple threads.

```
int i_epoch ;              // Each epoch is a complete pass through all training data
int n_threads ;            // Each batch is broken into this many threads
int ivis ;                 // Index within visible layer
int ihid ;                 // Index of hidden neuron
int istart ;               // Index in dataset of first batch case
int istop ;                // And one past last batch case
int jstart ;               // Offset in batch of first thread case
int jstop ;                // And one past last thread case
int n_in_batch ;           // Number of training cases in the batch being processed
int n_in_thread ;          // Number of training cases in the thread being processed
int ibatch ;               // Batch number being processed
int ithread ;              // Thread number being processed
int n_done ;               // Number of training cases done in this epoch so far
int nt_done ;              // Number of training cases done in this batch so far
int n_no_improvement ;     // N of consecutive times convergence crit failed to improve
double chain_length ;      // Chain length, which may be smoothed upwards
double error ;             // Mean squared error for each epoch; sum of squared diffs
double best_err ;          // Best error seen so far

int i, j, k, ret_val ;

double *dptr, momentum, max_inc, max_weight, error_vec[MAX_THREADS] ;
double best_crit, double sp_pen, x_this, x_prev, len_this, len_prev, dot ;
double smoothed_this, smoothed_dot, most_recent_correct_error ;

char msg[4096] ;

RBM_THR2_PARAMS params[MAX_THREADS] ;
HANDLE threads[MAX_THREADS] ;
```

The first step is to compute the mean vector of the inputs. This will be needed for the sparsity penalty. Then the data structure for passing information to the threads is initialized with those values that will not change. This code is shown here:

```
for (ivis=0 ; ivis<n_inputs ; ivis++)
  data_mean[ivis] = 0.0 ;

for (i=0 ; i<nc ; i++) {            // Pass through all cases, cumulating mean vector
  dptr = data + i * ncols ;        // Point to this case in the data
  for (ivis=0 ; ivis<n_inputs ; ivis++)
    data_mean[ivis] += dptr[ivis] ;
  }

for (ivis=0 ; ivis<n_inputs ; ivis++)
  data_mean[ivis] /= nc ;

for (i=0 ; i<max_threads ; i++) {
  params[i].mean_field = mean_field ;
  params[i].n_inputs = n_inputs ;
  params[i].ncols = ncols ;
  params[i].nhid = nhid ;
  params[i].data = data ;
  params[i].in_bias = in_bias ;
  params[i].hid_bias = hid_bias ;
  params[i].w = w ;
  params[i].shuffle_index = shuffle_index ;
  params[i].visible1 = visible1 + i * max_neurons ;
  params[i].visible2 = visible2 + i * max_neurons ;
  params[i].hidden1 = hidden1 + i * max_neurons ;
  params[i].hidden2 = hidden2 + i * max_neurons ;
  params[i].hidden_act = hidden_act + i * max_neurons ;
  params[i].in_bias_grad = in_bias_grad + i * max_neurons ;
  params[i].hid_bias_grad = hid_bias_grad + i * max_neurons ;
  params[i].hid_on_frac = hid_on_frac + i * max_neurons ;
  params[i].w_grad = w_grad + i * nhid * n_inputs ;
  params[i].error = error_vec + i ;
  }
```

In the previous code, the parameters in the first half are identical for all threads. The second half are work areas that must be maintained separately for each thread to avoid different threads stomping on one another's territory.

We'll be doing momentum smoothing of the parameter increments (Equation 3-23), so these increments must be initialized to zero. Also, the sparsity penalty will require a smoothed version of the fraction of time each hidden neuron is activated, hid_on_smoothed. We initialize this to 0.5.

```
for (ihid=0 ; ihid<nhid ; ihid++) {
  hid_bias_inc[ihid] = 0.0 ;
  hid_on_smoothed[ihid] = 0.5 ;
  for (ivis=0 ; ivis<n_inputs ; ivis++)
    w_inc[ihid*n_inputs+ivis] = 0.0 ;
  }

for (ivis=0 ; ivis<n_inputs ; ivis++)
  in_bias_inc[ivis] = 0.0 ;
```

We now do several more initializations: set up the shuffled index vector (page 118), zero the counter of contiguous failures to improve, and set the momentum and Markov chain length to their starting values. Then begin the loop that processes epochs (complete passes through the training set). The first step in each epoch is to shuffle the dataset indices.

```
for (i=0 ; i<nc ; i++)
  shuffle_index[i] = i ;

momentum = start_momentum ;
n_no_improvement = 0 ;          // Counts failure of ratio to improve
chain_length = n_chain_start ;

for (i_epoch=0 ; i_epoch<max_epochs ; i_epoch++) {

  i = nc ;                      // Number remaining to be shuffled
  while (i > 1) {               // While at least 2 left to shuffle
    j = (int) (unifrand_fast () * i) ;
    if (j >= i)
      j = i - 1 ;
```

```
    k = shuffle_index[--i] ;
    shuffle_index[i] = shuffle_index[j] ;
    shuffle_index[j] = k ;
    }
```

Within the epoch loop just begun, we have a batch loop. The training set is divided into batches, each of which is used to compute a gradient estimate and thereby update the weights.

```
istart = 0 ;              // Batch start = training data start
n_done = 0 ;             // Number of training cases done in this epoch so far
error = 0.0 ;             // Cumulates reproduction error
max_inc = 0.0 ;          // For testing convergence

for (ibatch=0 ; ibatch<n_batches ; ibatch++) {      // An epoch is split into batches
   n_in_batch = (nc - n_done) / (n_batches - ibatch) ; // Cases remaining / batches
   istop = istart + n_in_batch ;                        // Stop just before this index
```

At this point we are about to process a batch of n_in_batch cases beginning with case istart and ending just before case istop. We will break this batch into n_threads threads that run simultaneously. If the batch is large, use max_threads of them. But because of the overhead of starting a thread, we make sure each thread handles a fair number of cases. The constant 10, used here, is arbitrary and system dependent but not at all critical. Start all threads for this batch, first setting the parameters that vary.

```
n_threads = max_threads ;   // Try to use as many as possible
while (n_threads > 1 && n_in_batch / n_threads < 10) // But respect overhead
   --n_threads ;                    // The choice of constant is difficult

jstart = 0 ;                 // Thread within this batch
nt_done = 0 ;                // Number in this batch done

for (ithread=0 ; ithread<n_threads ; ithread++) {
   n_in_thread = (n_in_batch - nt_done) / (n_threads - ithread) ;
   jstop = jstart + n_in_thread ;

   params[ithread].istart = istart + jstart ;           // Batch start + thread start
   params[ithread].istop = istart + jstop ;             // Batch start + thread stop
   params[ithread].n_chain = (int) (chain_length + 0.5) ; // Constant for epoch
```

```
threads[ithread] = (HANDLE) _beginthreadex ( NULL , 0 , rbm2_wrapper ,
                                    &param s[ithread] , 0 , NULL ) ;
nt_done += n_in_thread ;    // Cases in this batch done so far
jstart = jstop ;            // Start the next thread where this thread stopped
} // For all threads in this batch
```

Now we just wait for all of the threads to finish. Recall from the initialization early in this routine that the gradient and error cumulation areas are offset for each thread by the amount of space needed. So, for example, params[0].hid_bias_grad contains the same address as hid_bias_grad and so forth for the other gradients and the error. Thus, we can start at thread 1 and sum the quantities for each thread into the root variable, which has the values for thread 0. This gives the net result for the entire batch. And we must close the thread handles.

```
ret_val = WaitForMultipleObjects ( n_threads , threads , TRUE , 1200000 ) ;

CloseHandle ( threads[0] ) ;
for (ithread=1 ; ithread<n_threads ; ithread++) { // Pool results into thread 0

   for (ihid=0 ; ihid<nhid ; ihid++) {
      hid_bias_grad[ihid] += (params[ithread].hid_bias_grad)[ihid] ;
      hid_on_frac[ihid] += (params[ithread].hid_on_frac)[ihid] ;
      for (ivis=0 ; ivis<n_inputs ; ivis++)
         w_grad[ihid*n_inputs+ivis] += (params[ithread].w_grad)[ihid*n_inputs+ivis] ;
      }

   for (ivis=0 ; ivis<n_inputs ; ivis++)
      in_bias_grad[ivis] += (params[ithread].in_bias_grad)[ivis] ;

   error_vec[0] += error_vec[ithread] ;
   CloseHandle ( threads[ithread] ) ;
   }
```

A batch has ended, and we have the gradient and reconstruction error for its pooled cases. We now perform the following actions:

- Cumulate in error the reconstruction error for this batch. We will sum this for the entire epoch.

- Divide hid_on_frac by the number of cases in this batch to get the average time each hidden neuron was activated in this batch.

- Exponentially smooth this quantity to get hid_on_smoothed.

- Use Equation 3-14 to compute sp_pen, the derivative of the sparsity penalty with respect to the net input of each neuron. Use the smoothed version of the mean activation just computed in order to prevent whipsawing.

- If the mean activation for a hidden neuron in this batch is very close to zero or one, impose an additional "sparsity" penalty to adjust the hidden bias and weights to pull the errant neuron away from useless saturation. The penalty factor of 0.5 is my heuristic and may safely be adjusted if desired.

- Use Equation 3-23 to compute the increment for the hidden bias. Since the input for the bias is fixed at one by definition, the derivative of the net input to the hidden neuron with respect to the bias is one; hence, the derivative of the sparsity penalty with respect to the bias is just sp_pen.

- Adjust the hidden bias per the increment.

This is done as follows:

```
error += error_vec[0] ;

for (ihid=0 ; ihid<nhid ; ihid++) {
   hid_on_frac[ihid] /= n_in_batch ;
   hid_on_smoothed[ihid] = 0.95*hid_on_smoothed[ihid] + 0.05*hid_on_frac[ihid] ;
   sp_pen = sparsity_penalty * (hid_on_smoothed[ihid] - sparsity_target) ;
   if (hid_on_frac[ihid] < 0.01)
      sp_pen += 0.5 * (hid_on_frac[ihid] - 0.01) ;     // 0.5 is heuristic
   if (hid_on_frac[ihid] > 0.99)
      sp_pen += 0.5 * (hid_on_frac[ihid] - 0.99) ;     // 0.5 is heuristic
   hid_bias_inc[ihid] = momentum * hid_bias_inc[ihid] +
               learning_rate * (hid_bias _grad[ihid] / n_in_batch - sp_pen) ;
   hid_bias[ihid] += hid_bias_inc[ihid] ;
```

Still within this hidden neuron loop, we handle the weight adjustments by doing the following in a visible neuron loop:

- Divide the gradient sum by the number of cases in the batch to get the mean gradient.

- Use Equation 2-19 to compute the weight penalty, which prevents weights from blowing up. This is also discussed on page 100. Adjust the gradient accordingly.

- We already computed sp_pen, the derivative of the sparsity penalty with respect to the input to each hidden neuron. The contribution to the input to a hidden neuron from a given visible neuron is the activation of that visible neuron times the weight connecting them. So, by the chain rule, the derivative of the sparsity penalty with respect to a weight is sp_pen for that hidden neuron times the activation of the visible neuron for that weight. Adjust the gradient per this quantity. Rather than trying to isolate each individual training case, it is easier and more stable to just use the mean of each input across the training set.

- Use Equation 3-23 to compute the increment for the weight and then adjust the weight by this quantity.

- Keep track of the maximum increment magnitude for the convergence test at the end of the epoch.

```
for (ivis=0 ; ivis<n_inputs ; ivis++) {
  w_grad[ihid*n_inputs+ivis] /= n_in_batch ;
  w_grad[ihid*n_inputs+ivis] -= weight_penalty * w[ihid*n_inputs+ivis] ;
  w_grad[ihid*n_inputs+ivis] -= data_mean[ivis] * sp_pen ;
  w_inc[ihid*n_inputs+ivis] = momentum * w_inc[ihid*n_inputs+ivis] +
                          learning_rate * w_grad[ihid*n_inputs +ivis] ;
  w[ihid*n_inputs+ivis] += w_inc[ihid*n_inputs+ivis] ;

  if (fabs(w_inc[ihid*n_inputs+ivis]) > max_inc) // For convergence test
    max_inc = fabs(w_inc[ihid*n_inputs+ivis]) ;
  } // For ivis
} // For ihid
```

Finally, we compute the increment for the visible bias and adjust accordingly.

```
for (ivis=0 ; ivis<n_inputs ; ivis++) {
   in_bias_inc[ivis] = momentum * in_bias_inc[ivis] +
                       learning_rate * in_bias _grad[ivis] / n_in_batch ;
   in_bias[ivis] += in_bias_inc[ivis] ;
   }
```

Now we can compute the angle (actually, its cosine) between the prior gradient and the current gradient in order to automatically adjust the learning rate and momentum, as discussed on page 118. If this is the first batch of the first epoch, there is no prior gradient, so initialize. Subsequently, we can compute the cosine of the angle, as shown here:

```
if (i_epoch == 0 && ibatch == 0) {
   len_this = 0.0 ;
   for (ihid=0 ; ihid<nhid ; ihid++) {
      for (ivis=0 ; ivis<n_inputs ; ivis++) {
         x_this = w_grad[ihid*n_inputs+ivis] ;
         w_prev[ihid*n_inputs+ivis] = x_this ;
         len_this += x_this * x_this ;
         }
      }
   len_prev = len_this ;
   }

else {
   len_this = dot = 0.0 ;
   for (ihid=0 ; ihid<nhid ; ihid++) {
      for (ivis=0 ; ivis<n_inputs ; ivis++) {
         x_this = w_grad[ihid*n_inputs+ivis] ;
         x_prev = w_prev[ihid*n_inputs+ivis] ;
         w_prev[ihid*n_inputs+ivis] = x_this ;
         len_this += x_this * x_this ;
         dot += x_this * x_prev ;
         }
      }

   dot /= sqrt ( len_this * len_prev ) ; // Cosine of angle between this and previous
   len_prev = len_this ;
```

With this angle cosine in hand, we can adjust the learning rate and momentum. The constants used here were heuristically determined from a few experiments and seem to work well. Readers should feel free to modify them as desired.

```
    if (dot > 0.5)
       learning_rate *= 1.2 ;
    else if (dot > 0.3)
       learning_rate *= 1.1 ;
    else if (dot < -0.5)
       learning_rate /= 1.2 ;
    else if (dot < -0.3)
       learning_rate /= 1.1 ;

    if (learning_rate > 1.0)
       learning_rate = 1.0 ;
    if (learning_rate < 0.001)
       learning_rate = 0.001 ;

    if (fabs(dot) > 0.3)
       momentum /= 1.5 ;
    } // Else we are in a subsequent batch
```

This is the end of the batch loop (within an epoch). Update the count of cases done in this batch and set up the starting index to begin the next batch where this one ended.

```
    n_done += n_in_batch ;
    istart = istop ;
    } // For each batch
```

We normalize the reconstruction error by dividing by the number of cases that went into the error sum as well as by the number of inputs. This gives us a mean error per case and input. Keep track of the best seen so far for the caller.

```
   error /= nc * n_inputs ;

  if (i_epoch == 0 || error < best_err)
    best_err = error ;
```

There are two tests for convergence, as described on page 120. The first shown here is secondary but important. It compares the maximum increment during the epoch to the maximum weight.

```
max_weight = 0.0 ;
for (ihid=0 ; ihid<nhid ; ihid++) {
  for (ivis=0 ; ivis<n_inputs ; ivis++) {
    if (fabs(w[ihid*n_inputs+ivis]) > max_weight)
      max_weight = fabs(w[ihid*n_inputs+ivis]) ;
    }
  }

if (max_inc / max_weight < convergence_crit)
  break ;
```

The primary convergence test is to count how many times in a row the ratio just computed fails to decrease. Numerous contiguous failures indicate that we are as close to a minimum as is practical to obtain.

```
if (i_epoch == 0 || max_inc / max_weight < best_crit) {
  best_crit = max_inc / max_weight ;
  n_no_improvement = 0 ; // Number of epochs with no improvement
  }

else {
  ++n_no_improvement ;
  if (n_no_improvement > max_no_improvement) // Test for convergence
    break ;
  }
```

That's almost it; the epoch is done. Slowly advance the momentum and Monte Carlo chain lengths toward their ending values.

```
momentum = 0.99 * momentum + 0.01 * end_momentum ;
chain_length = (1.0 - n_chain_rate) * chain_length + n_chain_rate * n_chain_end ;
```

One final bit of heuristic machination is done. In nearly all practical applications, when we are approaching convergence, the automatic learning rate adjustment already shown will do an excellent job; it will see that overshooting (a gradient reversing direction unless the steps are tiny) is a problem and hence will reduce the learning rate to a very small value. But in some pathological situations, generally associated with a lack of patterns in the training data, that algorithm produces learning rates that are somewhat too large. For this reason, when n_no_improvement tells us that we are nearly as low as we can get, it's good to force the learning rate to small values. In practice, this final heuristic will be invoked only rarely. But if the training data is essentially random, meaning that the RBM has little or nothing in the way of consistent patterns to model, the weights will bounce around for too long. No serious damage is done by this, other than delaying the completion of the algorithm. Still, delay is unpleasant, so this final heuristic can be useful.

```
    if (n_no_improvement > 50 && learning_rate > 0.03)
      learning_rate = 0.03 ;

    if (n_no_improvement > 100 && learning_rate > 0.02)
      learning_rate = 0.02 ;

    if (n_no_improvement > 150 && learning_rate > 0.01)
      learning_rate = 0.01 ;

    if (n_no_improvement > 200 && learning_rate > 0.005)
      learning_rate = 0.005 ;

    if (n_no_improvement > 250 && learning_rate > 0.002)
      learning_rate = 0.002 ;

    } // For each epoch

  return error ;
}
```

CUDA Code for RBM Training

Many principles of CUDA programming relevant to neural network training were discussed in the context of multiple-layer feedforward networks starting on page 36. You should review that material because the presentation of CUDA programming for RBMs will be much more terse and assume familiarity with previous material. Also, because complete CUDA source code for RBM training for free from the author's web site, not discuss all routines in this section. The discussion here will be limited to those routines that demonstrate important principles, are potentially confusing, or are unusual in some other way.

Initialization and Cache Line Matching

We already saw on page 50 that the first (L1) cache on CUDA devices is accessed through 128-byte reads and writes, and it is important to align global memory accesses in accordance with this as much as possible. Later, on page 51, we saw that there is a method for assuring perfect memory alignment, but that method was not used at that time because I felt that its relative complexity for MLFNs did not justify the trivial speed gain achieved, relative to the "almost perfect" method I used. So that the reader can see an example of this more complex technique, I employ it in the RBM training code where its inclusion is much easier than in an MLFN.

This technique applies to storage of matrices. The CUDA library guarantees that the address returned by a memory allocation routine is aligned with cache lines. But suppose a matrix contains a number of columns such that the number of bytes per row is not a multiple of 128 bytes? The first row will be perfectly aligned, but many subsequent rows will not be. This reduces the efficiency of global memory transfers.

The solution to this problem is to allocate extra space and pad the rows to contain a multiple of 128 bytes. This wastes scarce memory and makes the program a bit more complex, but it may increase speed enough to be worthwhile, though on modern CUDA devices the effect is often small.

We illustrate this process with several short code fragments. To facilitate efficient memory access when going from the visible layer to the hidden layer, as well as the reverse direction, we keep two copies of the weight matrix, one the transpose of the other. During initialization we extend the column dimensions of these matrices to a multiple of 128 bytes, which is 32 four-byte floats. Here is how these dimensions are computed, the memory allocated, and the weight matrix w on the host copied to the original matrix h_w and transpose h_wtr on the CUDA device. Note that we pad the unused memory at the end of each row with zeroes. We will see why this is necessary when we later discuss reduction algorithms. Array fdata is a temporary allocation for double-to-float conversion.

```
n_inputs_cols = (n_inputs + 31) / 32 * 32 ; // Row length of original matrix
nhid_cols = (nhid + 31) / 32 * 32 ;         // And transposed matrix

fdata = (float *) MALLOC ( n_inputs_cols * nhid_cols * sizeof(float) ) ;

cudaMalloc ( (void **) &h_w , (size_t) (n_inputs_cols * nhid * sizeof(float)) ) ;
```

```
for (j=0 ; j<nhid ; j++) {
  for (i=0 ; i<n_inputs ; i++)
    fdata[j*n_inputs_cols+i] = (float) w[j*n_inputs+i] ;
  for ( ; i<n_inputs_cols ; i++)
    fdata[j*n_inputs_cols+i] = 0.0f ;
  }

cudaMemcpy ( h_w , fdata , n_inputs_cols * nhid * sizeof(float) ,
               cudaMemcpyHostToDevice ) ;

cudaMalloc ( (void **) &h_wtr , (size_t) (n_inputs * nhid_cols * sizeof(float)) ) ;

for (i=0 ; i<n_inputs ; i++) {
  for (j=0 ; j<nhid ; j++)
    fdata[i*nhid_cols+j] = (float) w[j*n_inputs+i] ; // Transpose
  for ( ; j<nhid_cols ; j++)
    fdata[i*nhid_cols+j] = 0.0f ;
  }

cudaMemcpy ( h_wtr , fdata , n_inputs * nhid_cols * sizeof(float) ,
               cudaMemcpyHostToDevice ) ;
```

Fetching Training Cases

This section presents the algorithm that fetches the training data from d_data into d_visible1, shuffled and selected according to the batch index limits. If what you see is not clear, please review the material on MLFN CUDA training that began on page 36. Except for row padding, the addressing scheme used here is practically identical to that in the MLFN section. A code fragment from the host routine that invokes the device code for fetching appears first, followed by the device routine.

```
threads_per_block = (n_inputs + warpsize - 1) / warpsize * warpsize ;
if (threads_per_block > 4 * warpsize)
  threads_per_block = 4 * warpsize ;
block_launch.x = (n_inputs + threads_per_block - 1) / threads_per_block ;
block_launch.y = istop - istart ;
block_launch.z = 1 ;
```

```
device_fetch_vis1 <<< block_launch , threads_per_block >>> ( istart ) ;

__global__ void device_fetch_vis1 (
  int istart // First case in this batch
  )
{
  int icase, ivis ;

  ivis = blockIdx.x * blockDim.x + threadIdx.x ;
  if (ivis >= d_n_inputs)
    return ;

  icase = blockIdx.y ;

  d_visible1[icase*d_n_inputs_cols+ivis] =
                          d_data[d_s huffle_index[istart+icase]*d_n_inputs+ivis] ;
}
```

A key point to observe here is that the column multiplier for d_visible1 is d_n_inputs_ cols because this array has rows whose length is a multiple of the cache line width for maximum transfer efficiency. The multiplier for d_data is d_n_inputs because this row padding is not done for the training data.

The implication of not padding the training data matrix is that memory transfers from it will be less efficient than they would be if padding were done. However, this is not a serious problem. Data transfer from the training data matrix to d_visible1 happens only once per batch, while accesses to d_visible1 repeat during processing of a batch. The potentially large increase in storage space required to pad every row (case) of the training set does not justify the insignificant amount of time saved by padding.

Visible-to-Hidden Layer

This section discusses the CUDA code for computing hidden-layer probabilities from visible activations. We won't bother showing the host code that invokes it because it uses the same method as the code in the prior section. Each thread will handle a single hidden neuron.

```
__global__ void device_vis_to_hid (
  int nc              // Number of cases in this batch
  )
{
  int icase, ivis, ihid ;
  float sum, Q ;

  ihid = blockIdx.x * blockDim.x + threadIdx.x ;
  if (ihid >= d_nhid)
    return ;

  icase = blockIdx.y ;

  sum = d_hid_bias[ihid] ;
  for (ivis=0 ; ivis<d_n_inputs ; ivis++)
    sum += d_wtr[ivis*d_nhid_cols+ihid] * d_visible1[icase*d_n_inputs_cols+ivis] ;
  Q = 1.0f / (1.0f + __expf(-sum)) ;
  d_hidden1[icase*d_nhid_cols+ihid] = Q ;
  d_hidden2[icase*d_nhid_cols+ihid] = Q ;    // We'll need this for MC chain loop
  d_hid_on_frac[icase*d_nhid_cols+ihid] = Q ;
}
```

The big issue here, as usual, is memory access. We must use the transposed version of the weight matrix, d_wtr, so that the hidden neuron index changes the fastest in the matrix. This way, adjacent threads in a warp access adjacent elements in the weight matrix. The same access pattern happens in the three last lines, which store results. At first glance, d_visible1 may seem problematic because ivis changes the fastest. But with each pass through the loop, ivis will be constant for all threads, so a single fetch is sufficient to serve all threads! The hidden neuron index ihid doesn't even appear in addressing d_visible1.

Hidden-to-Visible Layer

The routine for bouncing back to the visible layer is slightly more complex because we have to deal with optional sampling of the visible activation. We use the nontransposed weight matrix for correct memory coalescing, and note that d_hidden_act has the same index for all threads in a warp, so memory issues are all taken care of.

The trick is the random number for sampling. We take advantage of the fact that d_shuffle_index is random. By combining the case index, the visible neuron index, and a random offset provided by the host caller, we can get a random integer ranging from zero through n_cases-1. Cute!

```
__global__ void device_hid_to_vis (
  int nc ,                    // Number of cases in this batch
  int random_offset          // Starting index in shuffle_index for random sampling
  )
{
  int k, icase, ivis, ihid ;
  float sum, P, frand ;

  ivis = blockIdx.x * blockDim.x + threadIdx.x ;
  if (ivis >= d_n_inputs)
    return ;

  icase = blockIdx.y ;

  sum = d_in_bias[ivis] ;
  for (ihid=0 ; ihid<d_nhid ; ihid++)
    sum += d_w[ihid*d_n_inputs_cols+ivis] * d_hidden_act[icase*d_nhid_cols+ihid] ;
  P = 1.0f / (1.0f + __expf(-sum)) ;

  if (d_mean_field)
    d_visible2[icase*d_n_inputs_cols+ivis] = P ;
  else {
    k = ((unsigned int) (icase * d_n_inputs + ivis + random_offset)) % d_ncases ;
    frand = (float) d_shuffle_index[k] / (float) d_ncases ;
    d_visible2[icase*d_n_inputs_cols+ivis] = (frand < P) ? 1.0f : 0.0f ;
    }
}
```

Gradient Length and Dot Product by Reduction

Reduction was covered in great detail on page 65, so we won't repeat that material. But the device code is listed here to show the reader why, when we copied the weights from the host to the device on page 142 and when we allocated the gradient, we had to pad the unused end of each row with zeros. This reduction algorithm, as well as one not shown that finds the maximum weight, string out the matrices as vectors, which leaves the unused ends interspersed throughout the vector!

```
__global__ void device_len_dot ()
{
  __shared__ float partial_len[REDUC_THREADS], partial_dot[REDUC_THREADS] ;
  int i, n, index ;
  float sum_len, sum_dot ;

  index = threadIdx.x ;
  n = d_n_inputs_cols * d_nhid ; // Includes unused padding at end of each row!
  sum_len = sum_dot = 0.0f ;
  for (i=blockIdx.x*blockDim.x+index ; i<n ; i+=blockDim.x*gridDim.x) { // String it out
    sum_len += d_w_grad[i] * d_w_grad[i] ;
    sum_dot += d_w_grad[i] * d_prev_grad[i] ;
    d_prev_grad[i] = d_w_grad[i] ;
    }

  partial_len[index] = sum_len ;
  partial_dot[index] = sum_dot ;
  __syncthreads() ;

  for (i=blockDim.x>>1 ; i ; i>>=1) {
    if (index < i) {
      partial_len[index] += partial_len[index+i] ;
      partial_dot[index] += partial_dot[index+i] ;
      }
    __syncthreads() ;
    }
  if (index == 0) {
    d_len_out[blockIdx.x] = partial_len[0] ;
    d_dot_out[blockIdx.x] = partial_dot[0] ;
    }
}
```

Updating the Input Bias

Updating the input bias is easy, but we'll list it here as an introduction to the more complex tasks of updating the hidden bias and weights. We use one thread per visible neuron. Here is the code:

```
__global__ void device_update_in_bias (
  int nc ,                  // Number of cases in this batch
  float rate ,              // Learning rate
  float momentum            // Learning momentum
  )
{
  int icase, ivis ;
  float sum ;

  ivis = blockIdx.x * blockDim.x + threadIdx.x ;

  if (ivis >= d_n_inputs)
    return ;

  sum = 0.0f ;

  for (icase=0 ; icase<nc ; icase++)
    sum += d_visible1[icase*d_n_inputs_cols+ivis] -
           d_visible2[icase*d_n_inputs_cols+ivis] ;

  d_in_bias_inc[ivis] = momentum * d_in_bias_inc[ivis] + rate * sum / nc ;
  d_in_bias[ivis] += d_in_bias_inc[ivis] ;
}
```

The most important thing to note is that every memory access is coalesced because adjacent memory locations are accessed by adjacent threads. The computation is just a straightforward application of Equation 3-12 and Equation 3-23.

Updating the Hidden Neuron Bias

The hidden neuron bias is more complex than the input bias because a sparsity penalty applies and also because random sampling may be required if a mean field approximation is not used. Here is the device code, and a discussion follows. We use one thread per hidden neuron.

```
__global__ void device_update_hid_bias (
  int nc ,                   // Number of cases in this batch
  float rate ,               // Learning rate
  float momentum ,           // Learning momentum
  int random_offset ,        // Starting index in shuffle_index for random sampling
  float sparse_pen ,         // Sparsity penalty
  float sparse_targ          // Sparsity target
  )
{
  int icase, ihid, k ;
  float sum, frac_on, frand ;

  ihid = blockIdx.x * blockDim.x + threadIdx.x ;
  if (ihid >= d_nhid)
    return ;

  sum = frac_on = 0.0f ;
  if (d_mean_field) {
    for (icase=0 ; icase<nc ; icase++) {    // Sum is Equation 3-12
      sum += d_hidden1[icase*d_nhid_cols+ihid] - d_hidden2[icase*d_nhid_cols+ihid] ;
      frac_on += d_hid_on_frac[icase*d_nhid_cols+ihid] ;  // Probability of being on
      }
    }
  else {
    for (icase=0 ; icase<nc ; icase++) {
      k = ((unsigned int) (icase * d_nhid + ihid + random_offset)) % d_ncases ;
      frand = (float) d_shuffle_index[k] / (float) d_ncases ;
      d_hidden_act[icase*d_nhid_cols+ihid] =
                        (frand < d_hidden1[ic ase*d_nhid_cols+ihid]) ? 1.0f : 0.0f ;
```

```
    sum += d_hidden_act[icase*d_nhid_cols+ihid] -
            d_hidden2[icase*d_nhid_cols+ihid] ;
    frac_on += d_hid_on_frac[icase*d_nhid_cols+ihid] ;
      }
    }

  sum /= nc ;
  frac_on /= nc ;

  d_hid_on_smoothed[ihid] = 0.95f * d_hid_on_smoothed[ihid] + 0.05f * frac_on ;
  sum -= sparse_pen * (d_hid_on_smoothed[ihid] - sparse_targ) ;
  if (d_hid_on_smoothed[ihid] < 0.01)
    sum -= 0.5 * (d_hid_on_smoothed[ihid] - 0.01) ;      // 0.5 is heuristic
  if (d_hid_on_smoothed[ihid] > 0.99)
    sum -= 0.5 * (d_hid_on_smoothed[ihid] - 0.99) ;

  d_hid_bias_inc[ihid] = momentum * d_hid_bias_inc[ihid] + rate * sum ;
  d_hid_bias[ihid] += d_hid_bias_inc[ihid] ;
}
```

First we sum the negative gradient according to Equation 3-12. If the user requests a mean field approximation (the generally preferable approach because it reduces random variation), then we can plug in the original (from the data) activation probabilities directly. But if mean field approximation is not desired, then we must sample that activation to get hidden_act. As we did earlier, we combine the case index, hidden neuron index, and a random offset supplied by the host caller to get a random integer from zero through d_n_cases-1 from the shuffle index array. This then lets us sample the hidden activation. In the same loop, we sum the fraction of time (probability) each hidden neuron was turned on by the input. This is needed for the sparsity penalty.

We exponentially smooth this fraction and plug it into Equation 3-14 to get the user's sparsity penalty gradient adjustment. The "visible activation" for a bias term is 1.0 by definition. We apply two more limited-effect "sparsity" penalties to prevent a hidden neuron from becoming saturated on or off. Finally, we use Equation 3-23 to adjust the hidden bias.

Updating the Weights

The final routine presented is that for updating the weight matrix. Here is the device code. A description will follow.

```
__global__ void device_update_weights (
  int nc ,                    // Number of cases in this batch
  float rate ,                // Learning rate
  float momentum ,            // Learning momentum
  float weight_pen ,          // Weight penalty
  float sparse_pen ,          // Sparsity penalty
  float sparse_targ           // Sparsity target
  )
{
  int icase, ivis, ihid ;
  float sum ;

  ivis = blockIdx.x * blockDim.x + threadIdx.x ;
  if (ivis >= d_n_inputs)
    return ;

  ihid = blockIdx.y ;

  sum = 0.0f ;
  if (d_mean_field) {
    for (icase=0 ; icase<nc ; icase++)
      sum += d_hidden1[icase*d_nhid_cols+ihid] *
             d_visible1[icase*d_n_inputs_cols+ivis] -
             d_hidden2[icase*d_nhid_cols+ihid] *
             d_visible2[icase*d_n_inputs_cols+ivis] ;
    }

  else {
    for (icase=0 ; icase<nc ; icase++)
      sum += d_hidden_act[icase*d_nhid_cols+ihid] *
             d_visible1[icase*d_n_inputs_cols+ivis] -
             d_hidden2[icase*d_nhid_cols+ihid] *
             d_visible2[icase*d_n_inputs_cols+ivis] ;
    }
```

```
   sum /= nc ;
   sum -= weight_pen * d_w[ihid*d_n_inputs_cols+ivis] ;
   sum -= d_data_mean[ivis] * sparse_pen * (d_hid_on_smoothed[ihid] - sparse_targ) ;
   if (d_hid_on_smoothed[ihid] < 0.01)
      sum -= d_data_mean[ivis] * 0.5 * (d_hid_on_smoothed[ihid] - 0.01) ; // 0.5 is heuristic
   if (d_hid_on_smoothed[ihid] > 0.99)
      sum -= d_data_mean[ivis] * 0.5 * (d_hid_on_smoothed[ihid] - 0.99) ;

   d_w_grad[ihid*d_n_inputs_cols+ivis] = sum ;
   d_w_inc[ihid*d_n_inputs_cols+ivis] =
                        momentum * d_w_inc[ihid*d_n_inputs+ivis] + rate * sum ;
   d_w[ihid*d_n_inputs_cols+ivis] += d_w_inc[ihid*d_n_inputs_cols+ivis] ;
}
```

First and most important, observe that the two variables that vary per thread, ihid and ivis, are defined such that the one that changes with contiguous threads, ivis, is also the one that defines contiguous memory locations in all arrays referenced in the routine. I know you are sick of hearing this, but such an arrangement is crucial to efficient memory access.

Astute readers will notice that I conveniently avoided updating d_wtr, the transposed weight matrix, in this routine. It would be addressed with these indices as d_wtr[ivis*d_nhid_cols+ihid]. As a result, contiguous threads, with their contiguous values of ivis, would have memory accesses that stride through memory with jumps of d_nhid_cols, causing very inefficient access. There are efficient ways to transpose a matrix using shared memory, so my original plan was to use such a routine to transpose the weight matrix after updates. But I programmed a naive transpose algorithm and, after a timing study, discovered that it ran in a tiny fraction of the time taken by the update, largely because the update requires math-intensive looping through the cases in the batch. So, I left my original design in place, but I do the transpose with a harmlessly inefficient naive algorithm, which I won't even bother listing here.

We sum the negative gradient according to Equation 3-12. Notice that if the user does not request a mean field approximation, d_hidden_act is used in the gradient computation, as was the case for the hidden neuron bias update. But we already computed this activation vector in the hidden bias routine, so there is no need to sample again.

The user's weight penalty, as described on page 100, is applied to the gradient.

The user's sparsity penalty, along with the same two additional penalties to avoid saturation, are applied just as they were for the hidden neuron bias in the prior section. This is a straightforward application of Equation 3-14.

Finally, we use Equation 3-23 to adjust the weights.

Putting It All Together

We have shown some of the most important CUDA routines for computing essential components of RBM training. The complete set of all routines is available for free download from the author's web site. In this section, we will show how these components can be assembled for CUDA-based RBM training. Here is the calling parameter list. All of these items have been seen before, so they will not be explained here.

```
double rbm_cuda (
    int nc ,                      // Number of cases in complete dataset
    int ncols ,                   // Number of columns in data
    double *data ,                // Nc rows by ncols columns of input data
    int n_inputs ,                // Number of inputs
    int nhid ,                    // Number of hidden neurons
    int n_chain_start ,           // Starting length of Markov chain, generally 1
    int n_chain_end ,             // Ending length of Markov chain
    double n_chain_rate ,         // Exponential smoothing rate for epochs
    int mean_field ,              // Use mean field instead of random sampling?
    int n_batches ,               // Number of batches per epoch
    int max_epochs ,              // Maximum number of epochs
    int max_no_imp ,              // Converged if this many epochs with no ratio imp
    double convergence_crit ,     // Convergence criterion for max inc / max weight
    double learning_rate ,        // Learning rate
    double start_momentum ,       // Learning momentum start value
    double end_momentum ,         // Learning momentum end value
    double weight_pen ,           // Weight penalty
    double sparsity_penalty ,     // Sparsity penalty
    double sparsity_target ,      // Sparsity target
    double *w ,                   // Computed weight matrix, nhid sets of n_inputs wts
    double *in_bias ,             // Computed input bias vector
    double *hid_bias ,            // Computed hidden bias vector
```

```
  int *shuffle_index ,              // Work vector nc long
  double *data_mean ,               // Work vector n_inputs long
  double *err_vec                   // Work vector n_inputs long
  )
```

The local variable declarations are shown next. Initialize the random number seed to a legal value and compute the mean of the inputs, which will be needed for sparsity penalties. Also initialize the shuffle index vector. For each epoch we will shuffle this and recopy it to the device for random batch selection as well as a random sampling of activations.

```
{
  int i, j, k, i_epoch, icase, ivis, n_no_improvement, ret_val, timer ;
  int istart, istop, ibatch, n_done, n_in_batch, max_batch, ichain, randnum ;
  double error, best_err, max_inc, momentum, chain_length ;
  double dtemp, sum, len_this, len_prev, dot, smoothed_this ;
  double smoothed_dot, max_weight, best_crit, most_recent_correct_error ;
  char msg[256] ;

  randnum = 1 ;

  for (ivis=0 ; ivis<n_inputs ; ivis++)
    data_mean[ivis] = 0.0 ;

  for (icase=0 ; icase<nc ; icase++) {
    for (ivis=0 ; ivis<n_inputs ; ivis++)
      data_mean[ivis] += data[icase*ncols+ivis] ;
    }

  for (ivis=0 ; ivis<n_inputs ; ivis++) {
    data_mean[ivis] /= nc ;
    if (data_mean[ivis] < 1.e-8)
      data_mean[ivis] = 1.e-8 ;
    if (data_mean[ivis] > 1.0 - 1.e-8)
      data_mean[ivis] = 1.0 - 1.e-8 ;
    }

  for (icase=0 ; icase<nc ; icase++)
    shuffle_index[icase] = icase ;
```

CUDA initialization will require the maximum batch size so that memory allocation can be performed correctly. So, we perform the batch loop that will be used later and keep track of the maximum size. We pass a character array to the initialization for it to pass back an error message if something goes wrong.

```
n_done = max_batch = 0 ;
for (ibatch=0 ; ibatch<n_batches ; ibatch++) {
  n_in_batch = (nc - n_done) / (n_batches - ibatch) ; // Cases left / batches left to do
  if (n_in_batch > max_batch)
    max_batch = n_in_batch ;
  n_done += n_in_batch ;
  }

ret_val = rbm_cuda_init ( nc , ncols , n_inputs , nhid , mean_field , max_batch , data ,
                          data_mean , in_bias , hid_bias , w , msg ) ;
```

We now begin the epoch loop, with each pass through this loop processing every case in the training set. Each epoch will be broken into batches for parameter update. Before beginning the epoch loop, we initialize the momentum and Markov chain length to their starting values. We also initialize to zero the counter of contiguous failures to improve. This will be your main convergence criterion. The first step of an epoch is to shuffle the index vector for random case selection. We then send this vector to the CUDA device.

```
momentum = start_momentum ;
chain_length = n_chain_start ;
n_no_improvement = 0 ; // Counts failure of ratio to improve

for (i_epoch=0 ; i_epoch<max_epochs ; i_epoch++) {

  i = nc ;                    // Number remaining to be shuffled
  while (i > 1) {             // While at least 2 left to shuffle
    j = (int) (unifrand_fast () * i) ;
    if (j >= i)               // This should never happen, but avoid disaster
      j = i - 1 ;
    k = shuffle_index[--i] ;
    shuffle_index[i] = shuffle_index[j] ;
    shuffle_index[j] = k ;
    }

  cuda_shuffle_to_device ( nc , shuffle_index ) ;
```

We now begin the batch loop within this epoch. Error will cumulate the reconstruction error across the epoch (all batches).

```
istart = 0 ;              // Batch start = training data start
n_done = 0 ;              // Number of training cases done in this epoch so far
error = 0.0 ;             // Cumulates reconstruction error across epoch
max_inc = 0.0 ;           // For testing convergence: increment relative to max weight

for (ibatch=0 ; ibatch<n_batches ; ibatch++) {       // An epoch is split into batches
  n_in_batch = (nc - n_done) / (n_batches - ibatch) ;  // Cases left to do / batches left
  istop = istart + n_in_batch ;                        // Stop just before this index

  // Get visible1 from data array
  cuda_fetch_vis1 ( istart , istop , n_inputs , NULL ) ;

  // Compute hidden1 probability (no sampling); also copy to hidden2 for MC chain
  cuda_vis_to_hid ( n_in_batch , nhid , NULL , NULL , NULL ) ;

for (ichain=0 ; ichain<(int)(chain_length+0.5) ; ichain++) { // Markov chain

  // Sample hidden2 into hidden_act
  k = randnum / IQ ;
  randnum = IA * (randnum - k * IQ) - IR * k ;
  if (randnum < 0)
    randnum += IM ;

  cuda_sample_hidden2 ( n_in_batch , nhid , randnum , NULL ) ;

  // Use hidden_act to get visible2, sampling visible2 if not mean_field
  k = randnum / IQ ;
  randnum = IA * (randnum - k * IQ) - IR * k ;
  if (randnum < 0)
    randnum += IM ;

  cuda_hid_to_vis ( n_in_batch , n_inputs , randnum , NULL ) ;

  if (ichain == 0) // Cumulate reconstruction error
    cuda_recon_error ( n_inputs , n_in_batch , err_vec ) ;
```

```
// Use visible2 (which is probabilities or samples per mean_field)
// to get hidden2 probabilities (no sampling of hidden2)
ret_val = cuda_vis2_to_hid2 ( n_in_batch , nhid , NULL ) ;
} // For Markov chain
```

The algorithm for computing the gradient just shown is given in outline form on page 99. We now continue that algorithm by updating the bias vectors and weight matrix. The random number is needed in case the user has not chosen a mean field approximation because the hidden probabilities under the data distribution must be sampled. After updating the weight matrix, save its transpose.

```
cuda_update_in_bias ( n_in_batch , n_inputs , learning_rate , momentum ,
                       NULL , NULL ) ;
k = randnum / IQ ;
randnum = IA * (randnum - k * IQ) - IR * k ;
if (randnum < 0)
  randnum += IM ;

cuda_update_hid_bias ( n_in_batch , nhid , learning_rate , momentum ,
                 randnum , sparsity_penalty , sparsity_target , NULL , NULL ) ;

cuda_update_weights ( n_in_batch , n_inputs , nhid , learning_rate ,
                      momentum , weight_pen , sparsity_penalty ,
                      sparsity_target , NULL , NULL , NULL ) ;

cuda_transpose ( n_inputs , nhid ) ;
```

We cumulate the reconstruction error across this epoch. Also keep track of the maximum weight increment so that we can use it for the secondary convergence criterion described on page 120.

```
for (ivis=0 ; ivis<n_inputs ; ivis++)
  error += err_vec[ivis] ; // Cumulates across epoch (all batches)

cuda_max_inc_w ( n_inputs * nhid , &dtemp , 1 ) ;
if (dtemp > max_inc)
  max_inc = dtemp ;
```

Compute the length of the weight gradient vector as well as its dot product with the prior gradient. If this is the very first evaluation, initialize.

```
if (i_epoch == 0 && ibatch == 0)
   cuda_len_dot ( n_inputs * nhid , &len_prev , &dot ) ;
```

We already have a previous gradient, so compute the dot product of the current with the previous and use the cosine of the angle separating the gradients to adjust the learning rate and momentum. The thresholds and adjustment factors are heuristic, so readers should feel free to experiment.

```
else {
   cuda_len_dot ( n_inputs * nhid , &len_this , &dot ) ;
   dot /= sqrt ( len_this * len_prev ) ;   // Cosine of angle
   len_prev = len_this ;

   if (dot > 0.5)      // Heuristic threshold
      learning_rate *= 1.2 ;
   else if (dot > 0.3)
      learning_rate *= 1.1 ;

   else if (dot < -0.5)
      learning_rate /= 1.2 ;
   else if (dot < -0.3)
      learning_rate /= 1.1 ;

   if (learning_rate > 1.0)
      learning_rate = 1.0 ;

   if (learning_rate < 0.001)
      learning_rate = 0.001 ;

   if (fabs(dot) > 0.3)
      momentum /= 1.5 ;
   }
```

This batch is complete. Tally the number of cases done so far in this epoch and set the starting case for the next batch to be the one right after the ending case for the current batch.

```
    n_done += n_in_batch ;
    istart = istop ;
    } // For ibatch
```

All batches of this epoch are complete. Normalize the reconstruction error to be per case and input. We already know the maximum increment across all batches, so now find the maximum weight. If the maximum increment is very small relative to the maximum weight, we are probably in a pathological situation of one or more weights blowing up, so quit. Or perhaps the user set the primary convergence criterion (checked next) to be unrealistically huge, so this ratio criterion gets us out.

```
    error /= nc * n_inputs ;
    cuda_max_inc_w ( n_inputs * nhid , &max_weight , 0 ) ;
    if (max_inc / max_weight < convergence_crit)
       break ;
```

Keep track of the smallest value of this ratio criterion. As long as it's regularly decreasing, either we are still moving toward the optimal parameter set (the usual situation) or a weight is blowing up (a rare pathological situation). The latter is handled by the secondary convergence criterion just presented. The normal situation is handled by counting how many times in a row we fail to improve. Once reasonable convergence is achieved, this count will rise. If not yet converged, adjust the momentum and Markov chain length. After convergence, fetch the parameters from the device and clean up (which frees memory on the device).

```
    if (i_epoch == 0 || max_inc / max_weight < best_crit) {
       best_crit = max_inc / max_weight ;
       n_no_improvement = 0 ; // Number of epochs with no improvement
       }
    else {
       ++n_no_improvement ;
       if (n_no_improvement > max_no_imp) // Test for convergence
          break ;
       }
```

```
    momentum = 0.99 * momentum + 0.01 * end_momentum ;
    chain_length = (1.0 - n_chain_rate) * chain_length + n_chain_rate * n_chain_end ;
    } // For i_epoch

  cuda_params_from_device ( n_inputs , nhid , in_bias , hid_bias , w ) ;
  rbm_cuda_cleanup () ;
}
```

On page 141 we saw a final heuristic for forcing a small learning rate in pathological situations involving nearly random training data. It looked like this:

```
  if (n_no_improvement > 50 && learning_rate > 0.03)
    learning_rate = 0.03 ;

  if (n_no_improvement > 100 && learning_rate > 0.02)
    learning_rate = 0.02 ;

  if (n_no_improvement > 150 && learning_rate > 0.01)
    learning_rate = 0.01 ;

  if (n_no_improvement > 200 && learning_rate > 0.005)
    learning_rate = 0.005 ;

  if (n_no_improvement > 250 && learning_rate > 0.002)
    learning_rate = 0.002 ;
```

This code block actually exists in the code shown on the prior page, occurring just before the end of the epoch loop, exactly as was the case in the threaded code of page 141. However, to keep end-of-training actions together and uncluttered, I omitted it from the prior page. If this is not clear, review that earlier reference, and examine the RBM_CUDA.CPP source file available for free download from the author's web site.

Timing

This section presents the comparative timing of the CUDA implementation just shown. To get an idea of the relative time of the various steps in the optimization, I used the 60,000-case 28×28 MNIST digit database with 400 hidden neurons, 10 epochs, and 100 batches per epoch. The performance of each step of the algorithm is shown in the following table. The first column of numbers is the total time for ten epochs, the second column is the time expressed as a percent of the total, and the last column is the time per kernel launch. In practice, this is a critical measure because most Windows systems automatically and without warning terminate a program that hits two seconds in a launch.

This information appears in a file called CUDA.LOG, which is automatically generated each time DEEP runs.

RBM CUDA times in seconds: total, (percent), per launch

Fetch batch data =	0.062	(0.4 percent)	0.000062 per launch
Visible to hidden1 =	2.478	(16.6 percent)	0.002478 per launch
Hidden to visible2 =	2.781	(18.6 percent)	0.001391 per launch
Visible2 to hidden2 =	2.434	(16.3 percent)	0.001217 per launch
Sample hidden2 =	0.094	(0.6 percent)	0.000094 per launch
Reconstruction =	0.232	(1.6 percent)	0.000232 per launch
Update input bias =	0.248	(1.7 percent)	0.000248 per launch
Update hidden bias =	0.282	(1.9 percent)	0.000282 per launch
Update weights =	5.928	(39.7 percent)	0.005928 per launch
Transpose =	0.062	(0.4 percent)	0.000062 per launch
Find max inc/w =	0.142	(1.0 percent)	0.000141 per launch
Dot product =	0.186	(1.2 percent)	0.000186 per launch

Updating Weights Analysis

Updating the weights is the biggest single eater of time, so a more detailed study will focus first on this routine. This study used the 60,000-case 28×28 MNIST digit database with 400 hidden neurons, 10 epochs, and 100 batches per epoch.

The first step in any CUDA analysis should be to check the occupancy, which (roughly) is the average ratio of the number of active warps to the maximum number that can be supported by the CUDA hardware. We want this to be as high as possible so that the warp scheduler can have numerous candidates from which to choose when one or more warps stall for any of various reasons discussed later.

The hardware-imposed limit on the number of active warps possible comes from a complex interaction between the number of blocks launched, the number of threads per block, shared memory usage, and register usage. The details are beyond the scope of this second, but we can take a look at the occupancy table shown in Figure 3-3. The *theoretical occupancy* is the upper limit attainable with your kernel and launch parameters. We should always strive to make this as close to 100 percent as possible, and this goal is attained here. The achieved occupancy is that which we obtain in real life, and execution inefficiencies almost never allow this to reach the theoretical limit. Anything over 50 percent is considered good, so 85.96 is great.

Variable	Achieved	Theoretical	Device Limit	
∧ Occupancy Per SM				
Active Blocks		16	16	
Active Warps	55.02	64	64	
Active Threads		2048	2048	
Occupancy	85.96 %	100.00 %	100.00 %	
∧ Warps				
Threads/Block		128	1024	
Warps/Block		4	32	
Block Limit		16	16	
∧ Registers				
Registers/Thread		24	255	
Registers/Block		3072	65536	
Registers/SM		49152	65536	
Block Limit		21	16	

Figure 3-3. *Weight update occupancy chart*

It is also beneficial to examine the way work is distributed among the multiprocessors. The workload should be balanced. If one or more of the multiprocessors are being assigned little work, the programmer has done a poor job of breaking the task into similar algorithms that can run in parallel and assigning those tasks to blocks and threads. The workload graphs are shown in Figure 3-4.

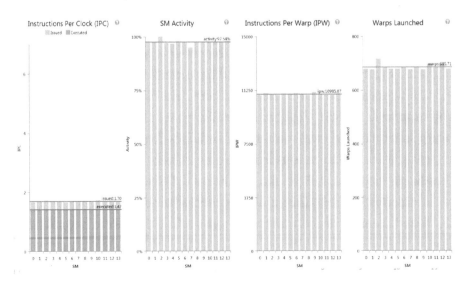

Figure 3-4. *Weight update occupancy chart*

Instructions Per Clock overlays the number of instructions issued and actually executed on each clock cycle. The quantity executed should be as close as possible to the quantity issued.

SM Activity is the percent of time each multiprocessor was active. If too few blocks are launched, some multiprocessors will finish their work and have nothing more queued up, resulting in wasted resources.

Instructions Per Warp is the average executed instructions per warp for each multiprocessor. This should be well balanced.

Warps Launched is the total number of warps launched per multiprocessor. Misbalance can happen when the program provided too few blocks in the launch. All of these weight update graphs are excellent.

Issue efficiency studies can reveal the subtle limitations imposed by the way the application is demanding hardware resources. The center chart in Figure 3-5 shows that for 63.65 percent of clock cycles, no warps were eligible for execution. This is not uncommonly high, but we need to find the cause for this stall. The right chart shows that the overwhelming reason is that a pipe was busy. Figure 3-6 shows that the arithmetic pipe is running nearly full-time, and 93.38 percent of that work is 32-bit integer arithmetic. The work of computing all those matrix subscripts is the limiting factor. Switching to direct pointer arithmetic might be faster, although pointers are 64 bits, so maybe not.

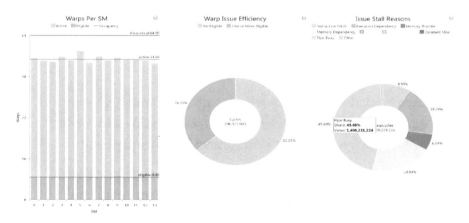

Figure 3-5. *Weight update issue efficiency and stall reasons*

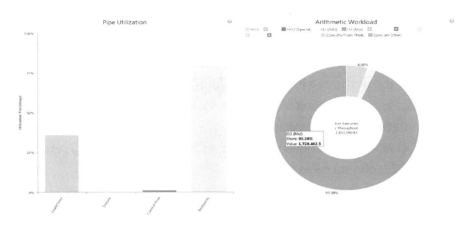

Figure 3-6. *Weight update pipe utilization*

Last but certainly not least, it always pays to examine line-by-line global memory access efficiency since this can be excruciatingly slow and will certainly be a limiting factor if accesses are done poorly. We've already beaten to death the idea that adjacent threads must, if at all possible, access adjacent memory locations, and to be perfect, the first thread in each warp should access a memory location that is an integer multiple of the cache line size, 128 bytes. We put a lot of effort into ensuring this, so it's worth checking to see if we succeeded. More than once I've noticed a suspicious memory efficiency statistic in the profiler and thereby found a serious bug in the program!

Look at Figure 3-7 and note the following items:

- At the very top we see that there are 6*400=2400 blocks launched, which is a goodly number and largely responsible for the excellent load balancing shown earlier.

- L1 Above Ideal Transactions is the number of memory transfer requests in which the ideal byte count was exceeded and hence transfer time was wasted. It's always zero, which is perfect.

- L1 Transfer Overhead is the ratio of the average number of bytes transferred to the number requested. Values greater than one indicate inefficient global memory access, and we have none.

- L2 Transfer Overhead is the similar statistic for the later, larger L2 cache. Curiously, two lines have values less than one, meaning that fewer bytes were transferred than requested. This comes about because these two lines of code have global references to memory whose address does not depend on ivis. As a result, the same quantity can sometimes be reused after by other threads in the warp after being fetched. This is a great bonus, but it's something that's difficult to plan for.

In summary, the memory access efficiency for this routine is excellent and about as perfect as it gets.

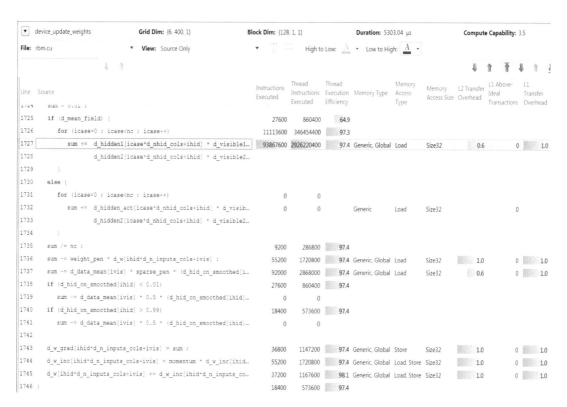

Figure 3-7. *Weight update memory access statistics*

Visible-to-Hidden Analysis

The second of the three fundamental time-eating RBM training algorithms is the one that computes the hidden layer from the visible layer. Its profiling charts are shown in Figures 3-8 through 3-12 without explanation to demonstrate its excellent behavior with the same task as used in the prior examples.

Variable	Achieved	Theoretical	Device Limit	
⌃ Occupancy Per SM				
Active Blocks		16	16	
Active Warps	44.35	64	64	
Active Threads		2048	2048	
Occupancy	69.30 %	100.00 %	100.00 %	
⌃ Warps				
Threads/Block		128	1024	
Warps/Block		4	32	
Block Limit		16	16	
⌃ Registers				
Registers/Thread		18	255	
Registers/Block		3072	65536	
Registers/SM		49152	65536	
Block Limit		21	16	

Figure 3-8. *Visible-to-hidden occupancy*

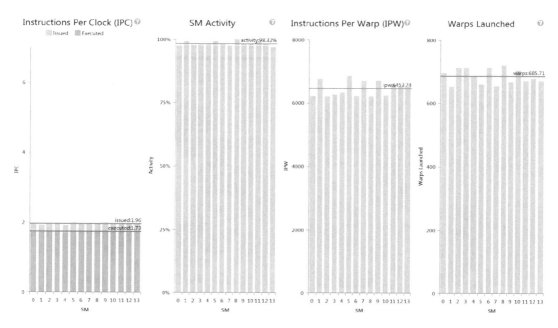

Figure 3-9. *Visible-to-hidden load balancing*

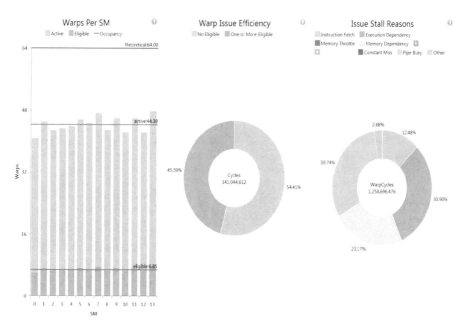

Figure 3-10. *Visible-to-hidden issue efficiency and stall reasons*

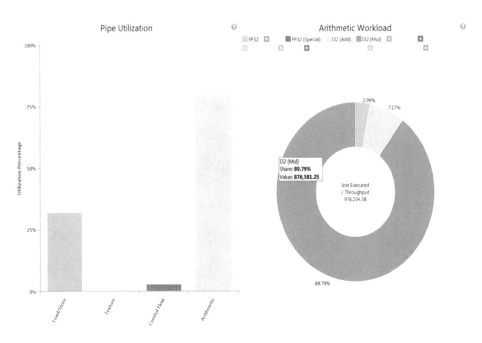

Figure 3-11. *Visible-to-hidden pipe utilization*

Figure 3-12. *Visible-to-hidden memory access statistics*

Hidden-to-Visible Analysis

The last of the three fundamental time-eating RBM training algorithms is the one that computes the visible layer from the hidden layer, also an excellent performer. The anomalous last line in the source code memory use listing is because of the optimizing compiler slightly confusing the profiler with its end-of-routine cleanup code (I checked the assembly output). See Figures 3-13 through 3-17.

Variable	Achieved	Theoretical	Device Limit	
⌄ Occupancy Per SM				
Active Blocks		16	16	
Active Warps	50.31	64	64	
Active Threads		2048	2048	
Occupancy	78.61 %	100.00 %	100.00 %	
⌄ Warps				
Threads/Block		128	1024	
Warps/Block		4	32	
Block Limit		16	16	
⌄ Registers				
Registers/Thread		18	255	
Registers/Block		3072	65536	
Registers/SM		49152	65536	
Block Limit		21	16	

Figure 3-13. *Hidden-to-visible occupancy*

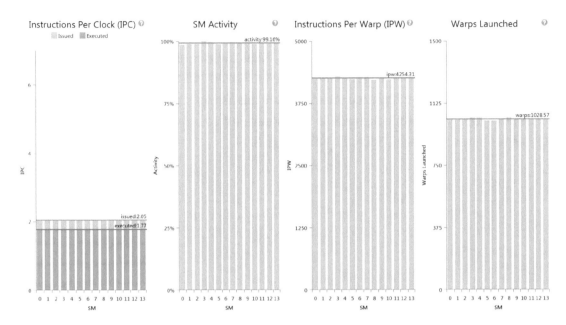

Figure 3-14. *Hidden-to-visible load balancing*

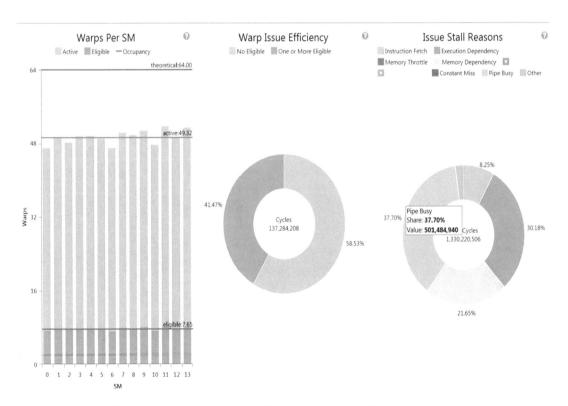

Figure 3-15. *Hidden-to-visible issue efficiency and stall reasons*

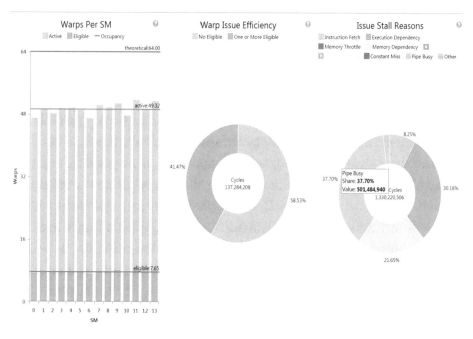

Figure 3-16. *Hidden-to-visible pipe utilization*

Line	Source	Instructions Executed	Thread Instructions Executed	Thread Execution Efficiency	Memory Type	Mem Access Type	Memory Access Size	L2 Transfer Overhead	L1 Above-Ideal Transactions	L1 Transfer Overhead
1049	int k, icase, ivis, ihid ;									
1050	float sum, P, frand ;									
1051										
1052	ivis = blockIdx.x * blockDim.x + threadIdx.x ;	43200	1382400	100.0						
1053	if (ivis >= d_n_inputs)	28800	921600	53.3						
1054	return ;									
1055										
1056	icase = blockIdx.y ;	13800	430200	97.4						
1057										
1058	sum = d_in_bias[ivis] ;	55200	1720800	97.4	Generic, Global	Load	Size32	1.0	0	1.0
1059	for (ihid=0 ; ihid<d_nhid ; ihid++)	11178000	348462000	97.1						
1060	sum += d_w[ihid*d_n_inputs_cols+ivis] * d_hidden_act[icase*d_...	49721400	1550010600	97.4	Generic, Global	Load	Size32	0.6	0	1.0
1061	P = 1.0f / (1.0f + __expf(-sum)) ;	27600	860400	97.4						
1062										
1063	if (d_mean_field)	27600	860400	48.7						
1064	d_visible2[icase*d_n_inputs_cols+ivis] = P ;	82800	2581200	97.4	Generic, Global	Store	Size32	1.0	0	1.0
1065	else {									
1066	k = ((unsigned int) (icase * d_n_inputs + ivis + random_offse...	0	0							
1067	frand = (float) d_shuffle_index[k] / (float) d_ncases ;	0	0		Generic	Load	Size32		0	
1068	d_visible2[icase*d_n_inputs_cols+ivis] = (frand < P) ? 1.0f...	14400	460800	100.0	Generic	Store	Size32		0	
1069	}									
1070										
1071	}	27600	860400	97.4						

Grid Dim: {6, 600, 1}   Block Dim: {128, 1, 1}   Duration: 2439.392 µs   Compute Capability: 3.5

File: rbm.cu   View: Source Only   High to Low: Low to High:

device_hid_to_vis

Figure 3-17. *Hidden-to-visible memory access statistics*

Advanced Training and Future Versions

Much work is being done in the development of training algorithms that run faster than the straight contrastive divergence algorithm shown here. These include particle methods, rapid weight decay methods, and various hybrid approaches. I am currently implementing some of these algorithms in threaded and CUDA versions and studying their behavior. I am also implementing auto-encoder algorithms as an alternative to RBMs.

There is an alternative CUDA programming approach to propagating activations from one layer to another, and this alternate algorithm is applicable to both supervised feedforward networks and RBMs. The CUDA implementations presented in this book have used the straightforward approach of distributing work according to individual neurons and carefully ensuring nearly perfect (for the feedforward nets) or perfectly perfect (for the RBMs) coalescing of global memory access. But propagation can also be seen as essentially a matrix-matrix multiplication followed by application of the activation function. When seen this way, one can implement highly efficient matrix multiplication using shared memory. I have a strong gut feeling that there is little to be gained with this vastly more complex method because arithmetic processing appears to be the limiting factor, with it completely hiding global accesses. However, I do intend to pursue this alternative to perform a rigorous comparison of the two methods.

My plan is to update the DEEP program and issue a supplement to this book when my development is complete. Stay tuned. Any announcements will be posted on the author's web site.

CHAPTER 4

Greedy Training

We saw in Chapter 2 how to construct and train (supervised) a traditional multiple-layer feedforward network. In Chapter 3 we learned how to construct and train without supervision a single-layer Restricted Boltzmann Machine (RBM). Now we will explore how to stack multiple RBMs into a deep belief network, train this multiple-layer network without supervision, and append an output layer that is trained with supervision. The resulting network can be enormously powerful.

The algorithm for stacking RBMs is surprisingly simple. Roughly stated, we train the bottom-most RBM, the one whose input is the training data. Once that's trained, we run the training cases through this model and use its hidden-layer activations as inputs to the next RBM, which we then train. When it is trained, we run the training data through the first and second RBMs and use the second's hidden activation as inputs for the third, which is then trained, and so on. This is why this training process is called *greedy*; no attempt is made to intelligently train the entire model at once. Rather, the training algorithm grabs as much as it can from the first layer alone and then locks it down and grabs as much as it can from the second layer, and so forth.

The traditional greedy training algorithm looks like this:

```
for (training_layer=0 ; training_layer<n_layers ; training_layer++) {    // Train this layer
   --> Initialize weights for training_layer

   While not converged { // Training loop                         |
      --> Get the raw input data from the database               |
      for (prop_layer=0 ; prop_layer<training_layer ; prop_layer++) {   |
         --> For each hidden neuron in prop_layer, compute probability  |
         --> Optionally sample hidden value using its probability       |
         } // Propagating through prior layers                   |
      --> Compute gradient, update weights for training_layer    |
      } // Training loop                                         |

   } // For all layers being trained
```

© Timothy Masters 2018
T. Masters, *Deep Belief Nets in C++ and CUDA C: Volume 1*, https://doi.org/10.1007/978-1-4842-3591-1_4

The lines marked with bars on the right are the training algorithm for the RBM being trained. There is nothing special about the algorithm itself; it can be the method shown in the prior chapter or whatever method the developer prefers. But a crucial difference is the fact that if we are sampling the hidden neurons at each stage, then the values used for computing the gradient for each batch change according to the vagaries of the random sampling. When we are training a single RBM or the bottom layer of a multiple-layer network, the inputs remain constant from batch to batch. But for layers past the first, the training inputs change.

This changing imposes a significant problem with the algorithm just shown when we sample the hidden neurons of each layer for presentation to the next layer as inputs. For every batch we must propagate the raw inputs through all layers prior to the one being trained. This is an expensive operation that adds considerable overhead to the training process.

For this reason I modified the prior algorithm so that if sampling is desired, it does so only for the single layer feeding the RBM being trained. All prior propagation is deterministic. This is a compromise that may trouble some experts, but it does allow random sampling yet without the overhead of constant repropagation. My algorithm is as follows:

```
for (training_layer=0 ; training_layer<n_layers ; training_layer++) {    // Train this layer
    --> Get the raw input data from the database
    for (prop_layer=0 ; prop_layer<training_layer ; prop_layer++)    // Propagate thru priors
        --> For each hidden neuron, compute probability for prop_layer
    --> Initialize weights for training_layer
    While not stopping criterion {                                  |
        --> Optionally sample value of feed from prior layer        |
        --> Compute gradient, update weights for training_layer     |
        } // Training this layer                                    |
    } // For training_layer; greedy training of all layers
```

As before, the lines marked on the right are the training algorithm. But now the only addition is the optional sampling of the hidden layer of the prior RBM that serves as input to the layer being trained. The propagation through all earlier layers is done just once before training begins, and those propagated values remain unchanged throughout all training batches.

Generative Sampling

One advantage of RBMs over other greedy model-building methods such as auto-encoding is that a trained RBM can be easily persuaded to reveal valuable insights about what it has learned. Recall from the introductory discussion that begins on page 5, as well as the insights given on page 93, that a properly trained RBM encapsulates the (mostly) authentic patterns inherent in the distribution represented by the training set. But what exactly are the patterns it is seeing? We can actually generate random samples from the distribution of patterns that it has encapsulated.

Now would be a good time to review page 91 and the several pages that follow. In particular, it must be clear that the distribution of random visible neuron states produced by the Markov chain of alternating Equations 3-1 and 3-2 mimics the distribution of cases in the training set. Naturally, contiguous samples from the chain would be substantially correlated. But samples from highly different starting vectors would have negligible correlation, especially as the randomness inherent in the sampling plays its role. The key point is that the trained RBM is a representation of the structure of the training data in the sense that it encapsulates the distribution of patterns inherent in the data.

No matter what visible or hidden set of activations one begins with, if you iterate between Equations 3-1 and 3-2 a very large number of times, the samples will eventually converge to the natural Markov chain implied by the RBM parameters, although perhaps tens of thousands of iterations will be required. If one happens to begin with an authentic sample from this distribution, the convergence will be instantaneous. The implication is that if the RBM has been well trained from a set of training data so that the RBM is a good representation of the data, then if one begins with a case sampled from the training set, convergence will usually be faster than if one begins with a totally random set of hidden or visible activations. This inspires us to frequently take this approach. Alternatively, one can begin with completely random activations and employ a very large number of iterations.

Things become slightly more complicated when one has constructed a deep belief network via the greedy training method described in the prior section. In this situation, we have one or more "trained and then fixed" layers that connect the raw input data at the bottom of the stack to the most recently trained layer at the top. This top layer is still an RBM. (Prior layers ceased being an RBM as soon as their weights were fixed and the layers were used to feed a subsequent RBM being trained.) Now we have to sample from this top-level RBM and then propagate its visible activations downward, all the way to the raw input layer at the bottom.

There are at least two ways to handle this situation. The most general way is to create a random activation vector for the RBM, run the Markov chain (often called *Gibbs sampling*) many times, and then propagate the final visible activations down to the bottom. This does have the disadvantage that it is unlikely that the random hidden-layer activations used to start the chain will be close to a legitimate sample from the RBM's distribution. Thus, a huge number of iterations will be required to achieve convergence to the Markov distribution. A frequently better approach is to randomly sample a training case and propagate it upward through the layers until it reaches the visible layer of the RBM sitting at the top of the stack. If the greedy training has been thorough, this set of visible activations will usually be close to the natural distribution of the RBM, and hence convergence will be fast.

A subroutine for performing this generative sampling is now shown. The calling parameter list is as follows:

```
static void gen_threaded (
   int nvis ,                   // Number of inputs to the first (bottom) layer
   int max_neurons ,            // Maximum number of neurons in any layer, as well as nvis
   int n_unsup ,                // Number of unsupervised layers
   int *nhid_unsup ,            // N_unsup vector containing the n of neurons in each layer
   double **weights_unsup ,// N_unsup pointers to weight matrices
   double *in_bias ,            // Input bias vectors; n_unsup sets of max_neurons each
   double *hid_bias ,           // Hidden bias vectors; n_unsup sets of max_neurons each
   int nchain ,                 // Length of Gibbs (Markov) chain, 0 to return raw data
   int input_vis ,              // Start with visible (as opposed to hidden)?
   double *workvec1 ,           // Work vector max_neurons long, also inputs starting case
   double *workvec2 ,           // Work vector max_neurons long, also inputs starting hidden
   unsigned char *image         // Computed image, 0-255 returned here
   )
```

There are nvis raw data inputs. The maximum number of neurons in any layer is max_neurons. There are n_unsup layers in the deep belief net, including the RBM. The vector nhid_unsup contains the number of hidden neurons in each layer, and the vector weights_unsup contains pointers to the weight matrix for each layer. The vectors in_bias and hid_bias contain the concatenated (all layers) input and hidden bias vectors, respectively. Each layer's bias vector is max_neurons long, although elements at the end will generally be unused. We will perform nchain Gibbs sampling (Markov chain) iterations, beginning with a visible activation vector input in workvec1 if input_vis is true or beginning with a hidden activation vector input in workvec2 if input_vis is false. The output of this subroutine is in image, with the values 0 through 255 according to the sampled input activations. If actual probabilities are desired, the modification is trivial and obvious. Finally, if nchain is input as zero, no sampling is done; the input activations are scaled 0 to 255 and returned in image.

The constant (for random numbers) and variable declarations are shown next. We will be switching back and forth between using workvec1 and workvec2 for the visible and hidden neuron activations. Initialize them. If the user wants no sampling, just rescale the inputs and return.

```
#define IA 16807
#define IM 2147483647
#define AM (1.0 / IM)
#define IQ 127773
#define IR 2836

  int i, k, ichain, ivis, nin, ihid, nhid, i_layer, randnum ;
  double *vis_layer, *hid_layer, *w, *wptr, *ibptr, *hbptr, sum, Q, frand ;

  vis_layer = workvec1 ;
  hid_layer = workvec2 ;

  if (nchain == 0) {   // User wants original image? This overrides input_vis.
    for (i=0 ; i<nvis ; i++)
      image[i] = (unsigned char) (255.9999 * vis_layer[i]) ;
    return ;
    }
```

If the user wants to start the chain with an input vector sample (typically a training case for rapid convergence), then we must propagate upward to the RBM that sits at the top. We'll need a random number generator for sampling hidden neuron activations in the Gibbs chain. Use a crude but effective method for initializing the seed.

```
  if (input_vis) {

    randnum = 1 ;            // Get a somewhat random seed
    for (i=0 ; i<nvis ; i++) {   // The seed must never be zero!
      if (vis_layer[i] > 0.5)
        ++randnum ;
      }
```

```
// Propagate up until we reach the RBM

  nin = nvis ;                            // Number of inputs to the current layer
  for (i_layer=0 ; i_layer<n_unsup-1 ; i_layer++) {
    nhid = nhid_unsup[i_layer] ;    // Number of hidden neurons in the current layer
    w = weights_unsup[i_layer] ;   // Its weight matrix
    hbptr = hid_bias + i_layer * max_neurons ;   // Hidden bias vector for this layer
      for (ihid=0 ; ihid<nhid ; ihid++) {       // Compute all hidden neuron activations
        wptr = w + ihid * nin ;                 // Weight vector for this neuron
        sum = hbptr[ihid] ;                     // This hidden neuron's bias
        for (ivis=0 ; ivis<nin ; ivis++)        // Equation 3-3
          sum += wptr[ivis] * vis_layer[ivis] ;
        hid_layer[ihid] = 1.0 / (1.0 + exp(-sum)) ;
        }
    nin = nhid ;                            // Hidden for this layer is 'visible' for next
    if (vis_layer == workvec1) {            // Switch back and forth
      vis_layer = workvec2 ;
      hid_layer = workvec1 ;
      }
    else {
      vis_layer = workvec1 ;
      hid_layer = workvec2 ;
      }
    } // For i_layer, propagating up until the RBM
  } // If input_vis
```

If, instead, the user is inputting a vector of (presumably random) hidden neuron activations for the RBM at the top (layer n_unsup-1), then we use those activations as a crude way to initialize the random seed. Also, we set the number of inputs to the RBM. If it happens to be the only layer, this is just the number of raw data inputs. But if there are one or more layers below the RBM, then its inputs are the hidden neurons for the layer just under it.

```
else { // Not input_vis, so user is inputting hidden layer of RBM
  randnum = 1 ;          // Get a somewhat random seed
  for (i=0 ; i<nhid_unsup[n_unsup-1] ; i++) {
    if (hid_layer[i] > 0.5)
      ++randnum ;
    }
```

```
    if (n_unsup == 1)          // If there are no layers below the RBM
      nin = nvis ;
    else
      nin = nhid_unsup[n_unsup-2] ;
    } // If not input_vis
```

We are now ready for the Markov chain, which alternates between visible-to-hidden activation using Equations 3-3 and 3-4. Recall that we *must* sample when computing the hidden activations. Sampling is optional when computing the visible activations, and here we avoid this sampling, which introduces an extra degree of randomness.

The first step is to set the number of hidden neurons of the RBM and get pointers to its weight matrix and bias vectors.

```
  nhid = nhid_unsup[n_unsup-1] ;
  w = weights_unsup[n_unsup-1] ;
  hbptr = hid_bias + (n_unsup-1) * max_neurons ;
  ibptr = in_bias + (n_unsup-1) * max_neurons ;
```

The Gibbs sampling (Markov iterations) chain is shown next. The first half of the loop computes the vector of hidden-layer activations, with sampling. The second half computes the visible layer activations, without sampling. Note that if the user has chosen to input hidden-layer activations as the starting point (input_vis false), then we must skip the first half of the chain loop for the first pass. We jump right in with the hidden-to-visible phase.

```
  for (ichain=0 ; ichain<nchain ; ichain++) {

    if (ichain || input_vis) {          // Skip first visible-to-hidden if user inputs hidden
      for (ihid=0 ; ihid<nhid ; ihid++) {   // Visible to hidden, with sampling
        wptr = w + ihid * nin ;             // Weight vector for this neuron
        sum = hbptr[ihid] ;                 // This hidden neuron's bias
        for (ivis=0 ; ivis<nin ; ivis++)    // Equation 3-3
          sum += wptr[ivis] * vis_layer[ivis] ;
        Q = 1.0 / (1.0 + exp(-sum)) ;
        k = randnum / IQ ;                 // Random generator for sampling
        randnum = IA * (randnum - k * IQ) - IR * k ;
```

```
      if (randnum < 0)
        randnum += IM ;
      frand = AM * randnum ;
      hid_layer[ihid] = (frand < Q) ? 1.0 : 0.0 ;  // Sample
      }
    }
  for (ivis=0 ; ivis<nin ; ivis++) {          // Hidden to visible, without sampling
    sum = ibptr[ivis] ;                        // Input bias
    for (ihid=0 ; ihid<nhid ; ihid++)          // Equation 3-4
      sum += w[ihid*nin+ivis] * hid_layer[ihid] ;
    vis_layer[ivis] = 1.0 / (1.0 + exp(-sum)) ;
    }

  } // For ichain
```

After the Gibbs sampling of the Markov chain is complete, we must propagate the RBM's visible layer down through prior layers until we reach the raw data input layer. As we did when propagating upward, we alternate using workvec1 and workvec2 for the visible and hidden layers.

```
for (i_layer=n_unsup-2 ; i_layer>=0 ; i_layer--) {
  nhid = nin ;             // Hidden neurons for this layer are RBM's visible neurons
  if (i_layer == 0)        // If we reached the bottom (raw data) layer
    nin = nvis ;           // Then n of 'inputs' for this layer are n of raw inputs
  else                     // Else they are hidden of the layer just below this one
    nin = nhid_unsup[i_layer-1] ;
  w = weights_unsup[i_layer] ;          // This layer's weight matrix
  ibptr = in_bias + i_layer * max_neurons ; // And input bias vector

  if (vis_layer == workvec1) {     // Swap so we always map to the 'other' vector
    vis_layer = workvec2 ;
    hid_layer = workvec1 ;
    }
  else {
    vis_layer = workvec1 ;
    hid_layer = workvec2 ;
    }
```

```
   for (ivis=0 ; ivis<nin ; ivis++) {              // Hidden to visible, without sampling
     sum = ibptr[ivis] ;
     for (ihid=0 ; ihid<nhid ; ihid++)           // Equation 3-4
       sum += w[ihid*nin+ivis] * hid_layer[ihid] ;
     vis_layer[ivis] = 1.0 / (1.0 + exp(-sum)) ;
     }
   } // For i_layer, propagating down to the data input layer
```

The final step is to map the 0–1 activations to the 0–255 image tones. Feel free to modify this step as desired.

```
  for (i=0 ; i<nvis ; i++)
    image[i] = (unsigned char) (255.9999 * vis_layer[i]) ;
}
```

CHAPTER 5

DEEP Operating Manual

This chapter presents a concise operating manual for DEEP 1.0, a Windows application. It describes every menu option and lists the page number where you can find more details if the short description here is not sufficient. (Version 2.0, which is fully backward compatible, is now available for free download.)

Menu Options

This section covers the menu options.

File Menu Options

These are the File menu options:

Read a database, page 186

> A text file in standard database format (such as Excel CSV) is read. The first line names the variables, and subsequent lines are the data, one case per line. Spaces, tabs, and commas can be used as delimiters. Subsequent training will produce a predictive model by default, not a classifier.

Read MNIST image, page 186

> A standard MNIST-format image file is read. The corresponding MNIST label file must be read after the image file is read. Subsequent training will produce a model that is a classifier by default, not a predictive model.

Read MNIST labels, page 187

> A standard MNIST-format label file is read. The corresponding MNIST image file must be read before the label file is read.

© Timothy Masters 2018
T. Masters, *Deep Belief Nets in C++ and CUDA C: Volume 1*, https://doi.org/10.1007/978-1-4842-3591-1_5

Write activation file, page 187

> A text file containing the activation of a specified neuron for all training set cases is written.

Clear all data, page 187

> All training data is erased, but a trained model (if it exists) is retained. The purpose of this command is to allow reading a test dataset and evaluating the performance of a trained model on this new dataset.

Print

> The currently selected display window (created under the Display menu) is printed. If no window is selected, Print is disabled.

Exit

> The program is terminated.

Test Menu Options

These are the Test menu options:

Use CUDA (Toggle Yes/No)

> This option is enabled only if a CUDA-capable device is present on the computer. If a check mark appears next to this option, the CUDA device will be used for compute-intensive operations. Click this option to toggle the check mark on and off.

Model Architecture, page 188

> The number of unsupervised and supervised layers is specified, as well as the number of neurons in each layer. If the data was read with the "Read a database" command, the model will be predictive by default, predicting numeric values of the target variables. If MNIST data was read, the model will be a classifier by default, employing a SoftMax output layer to classify according to the labels in the label file.

Database inputs and targets, page 189

> The user specifies one or more predictor variables and one or
> more target variables. If MNIST data was read, the predictors
> and targets are predefined and need not be specified by the
> user. However, the user can still change them through this menu
> command if desired. During model training, predictors that are
> constant for all training cases are omitted from the model.

Advanced options

> Options of an advanced nature and that would not normally be
> changed by the user can be set here. In DEEP 1.0 the only such
> option is the maximum number of threads allowed for non-CUDA
> threaded computation. The default should be excellent in all
> practical applications. It cannot be set to more than 64 because of
> limitations imposed by the Windows operating system.

RBM training params, page 189

> Parameters relevant to unsupervised RBM training can be set.

Supervised training params, page 193

> Parameters relevant to training the supervised layers can be set.

Train, page 196

> The complete model is trained using the data currently present.

Test, page 198

> The trained model is tested with the data currently present.

Analyze, page 200

> Two basic analyses of the trained model are performed. The first is
> a comparison of the mean activation of inputs compared to those
> for the reconstructed data, and the second is a mean activation of
> the final unsupervised layer.

Display Menu Options

These are the Display menu options:

Receptive field, page 201

> A plot of the receptive fields (weights of the first/bottom layer) for one or more hidden neurons is displayed. This display can be printed with the File ➤ Print command.

Generative sample, page 202

> A plot of one or more generative samples is displayed. This display can be printed with the File ➤ Print command.

The "Read a database" Option

A text file in standard database format is read. In particular, standard-format Excel CSV files may be read, as well as databases produced by many common statistical and data analysis programs. The first line must specify the names of the variables in the database. The maximum length of each variable name is 15 characters. The name must start with a letter and may contain only letters, numbers, and the underscore (_) character.

Subsequent lines contain the data, one case per line. Missing data is not allowed.

Spaces, tabs, and commas can be used as delimiters for the first (variable name) and subsequent lines.

Here are the first few lines from a typical data file. Six variables are present, and three cases are shown.

```
RAND0 RAND1 RAND2 RAND3 RAND4 RAND5
-0.82449359  0.25341070   0.30325535  -0.40908301  -0.10667177  0.73517430
-0.47731471 -0.13823473  -0.03947150   0.34984449   0.31303233  0.66533709
 0.12963752 -0.42903802   0.71724504   0.97796118  -0.23133837  0.81885117
```

The "Read MNIST image" Option

A standard MNIST image file is read. It is assumed that there will be ten labels. The number of rows and columns is read from the file and not assumed by DEEP, although the common file is 28 rows and columns. In DEEP 1.0, the product of the number of rows and columns must not exceed 4096–10=4086. There is no hard-coded limit on the number of images; it is limited only by available memory.

Models in DEEP 1.0 can be either classifiers, in which case the output layer is SoftMax, or predictive, in which case the output layer is linear with no range limiting, and it makes numeric predictions. When MNIST data is read, the classifier form of model is used by default. For database data, the default is numeric prediction. In both cases, the user can override the default and force the model to be a classifier or predictive.

The MNIST image file must be read before a label file can be read.

The "Read MNIST labels" Option

A standard MNIST label file is read. It is assumed that there are ten possible labels. The label file cannot be read until the image file has been read.

The "Write activation file" Option

This option writes a text file containing the activation of a single neuron for all cases, one line per case. The user specifies whether the neuron to be written is in the unsupervised or supervised section, which layer within that section it is in (with 1 being the first layer), and the neuron number within that layer (also with 1 being the first neuron).

An activation file is mainly for diagnostic use, although some users may find it convenient to pass an activation file to other programs.

The "Clear all data" Option

Sometimes the user will want to test a trained model on data that the model has not yet seen, often called a *test set* or *out-of-sample* (OOS) data. This can be done by reading the training data, training the model, clicking "Clear all data," reading the test set, and clicking Test.

When a trained model exists and data is cleared, subsequently read data must have the same variables in the same order as the data that was used to train the model.

Model Architecture

A model in DEEP 1.0 consists of zero or more unsupervised layers created by greedy RBM training, followed by one or more supervised layers trained by using the outputs of the final unsupervised layer (or the raw data if there are no unsupervised layers) as inputs and targets as outputs. The user defines the architecture by specifying the following quantities:

Number of unsupervised layers: This may be zero to create a model that is entirely supervised.

Hidden neurons in first unsupervised layer: This refers to the bottom layer, the one closest to the input data.

Hidden neurons in last unsupervised layer: This refers to the topmost RBM layer, the one that feeds the supervised section. If there is only one unsupervised layer, this must equal "Hidden neurons in first unsupervised layer." If there are multiple layers, interior sizes are linearly interpolated.

Number of supervised layers: This must be at least one (the output layer), which is the usual case when there are one or more unsupervised layers. But it is legal for an unsupervised RBM section to feed a "traditional" supervised model, one having one or more hidden layers prior to the output layer. It is also possible to use DEEP 1.0 for strictly supervised models.

Hidden neurons in first supervised layer: This is relevant only if "Number of supervised layers" is greater than one, in which case it is the number of hidden neurons in the first layer encountered by the unsupervised layer outputs or the raw data if there are no unsupervised layers.

Hidden neurons in last supervised layer: This refers to the last hidden layer before the output layer. If "Number of supervised layers" is two (one hidden, plus output), this must equal "Hidden neurons in first supervised layer." If there are multiple hidden layers ("Number of supervised layers" exceeds two), interior sizes are linearly interpolated.

Database Inputs and Targets

This option is used to specify the variables that will be used as inputs to the model (the *predictors*) and the variables that will be predicted (the *targets*). One or more of each can be selected using standard Windows methods: dragging across a range, holding down Shift while clicking the first and last in a range, or holding down Control to select individual variables.

If "Read a database" was used to read the training data, then the user must specify the inputs and targets. But if the data is an MNIST file set, then the inputs and targets are automatically preset. Nonetheless, the user is free to use this menu option to change the preset selection.

All MNIST input variables will follow the naming convention of P_*row_column* to identify the location of each pixel in the input grid, with the naming origin (first row/column) being zero. Thus, the upper-left pixel will be P_0_0.

The MNIST target variables will be named Label_*digit* to identify the digit with which each class is associated. Thus, the targets will be named Label_0 through Label_9.

For MNIST data, the model will be a classifier with SoftMax outputs by default. For training data read from a database, the model will by default be predictive, attempting to predict numeric values for each target. But a supervised training option (described later) allows the user to force the model to be a classifier or predictor. For a forced classifier, the user must specify at least two targets using the "Data inputs and target" menu option, and for each case, the target having the maximum value will be assumed to identify the class of the case.

RBM Training Params

This menu option sets the parameters that are relevant to RBM training. All parameters are preset to defaults that should be reasonable for many or most applications. The following parameters may be set:

> *Random initialization iterations*: This is the number of trial weight sets that are tested to find a good starting point for stochastic gradient descent training. This algorithm is described starting on page 103. It is definitely worthwhile doing at least a few dozen trials so that subsequent training begins with a reconstruction error that is not outrageous. More than several hundred trials is probably overkill.

Number of batches: The training set is divided into this many batches (though the exact number may be adjusted by the program when necessary) for stochastic gradient descent. Concepts vital to this choice are discussed beginning on page 117. Here are the basic principles:

- Recall from the cited discussion that the trade-off between time-per-batch and batches-for-convergence is unbalanced in the direction of favoring many small batches. But consider the next point.

- Although Windows threads have fairly small overhead, the overhead of launching a CUDA kernel can be considerable. Thus, one should be inclined to use fewer batches if using CUDA processing.

- The automatic learning rate and momentum adjustment algorithms described in the section starting on page 118 perform best with relatively large batches. This should inspire us to use few batches.

- Most important in practice, most Windows installations impose an upper limit of two seconds for a CUDA kernel, after which it is given the boot. Kernel time is almost linearly related to batch size, so if your screen blacks out and recovers with a message that the driver was reset, increase the number of batches. CUDA.LOG lists kernel times and hence can be used to see how close to criticality you are (see page 160).

Markov chain length (CD-k) start: When stochastic gradient descent begins, this is the number of iterations taken by executing the Markov chain in the contrastive divergence algorithm shown on page 99. The gradient estimate's accuracy is improved by taking more iterations, with the result that convergence requires fewer epochs. But these samples are expensive to obtain. Early in training we do not need accurate gradient estimates; a rough approximation is sufficient. This parameter should almost always be left at its default value of one.

Markov chain length (CD-k) end: This is the number of iterations taken as learning progresses. As convergence nears, it is worthwhile to expend computation time to obtain more accurate gradient estimates. The default value of four is good in nearly all applications. In case the user wants to obtain true maximum likelihood parameter estimates (usually pointless in practice), you can set this parameter to a very large value.

Markov chain length (CD-k) rate: This is the rate at which the chain length increases from the starting value to the end value. Standard exponential smoothing is employed, with the ending chain length being the "new value" of the series.

Learning rate: This is the initial learning rate. This should be small, probably smaller than the value the user is accustomed to for other programs. This is because the automatic adjustment algorithm described starting on page 118 will rapidly move it to an optimal value.

Momentum start: This is the initial momentum. This should be small, probably smaller than the value the user is accustomed to for other programs. As with the learning rate, the automatic adjustment algorithm described starting on page 118 will rapidly move it to an optimal value.

Momentum end: As training progresses, the momentum will progress toward this value unless the adjustment algorithm swats it down because of instability in the gradient descent algorithm. Values greater than the default are dangerous, and even the default is pretty high.

Weight penalty: This is the degree to which large weights are penalized. This must be small to allow weights to approach their optimal values. But it should not be zero. If no weight penalty is applied, in unusual but annoying pathological situations one or more weights can blow up to enormous values.

Sparsity penalty: This is the degree to which hidden neuron activation rates are encouraged to approach the *sparsity target* specified as the next parameter. This is not a critical parameter, and it can safely be set to zero if desired. However, in most cases it is good to gently nudge weights toward values that result in smallish hidden neuron activation rates, such as 0.1 or so. Among other things, this makes the weights more interpretable because one can then study which patterns are associated with activation of certain hidden neurons. If all hidden neurons are activated about half the time, such interpretation is more difficult than if activation is rarer.

Sparsity target: This is the value toward which hidden neuron activation rates are nudged by the *sparsity penalty*. This is typically around 0.1 or so. This parameter is ignored if the *sparsity penalty* is zero.

Increment convergence criterion: This is the secondary convergence criterion, as described starting on page 120. If the ratio of the magnitude of the largest weight adjustment in an epoch to the magnitude of the largest weight drops below this threshold, convergence is decreed to be complete. This should be small to avoid early exits from the training algorithm.

Max epochs with no improvement: This is the primary convergence criterion. The ratio of the magnitude of the largest weight adjustment in an epoch to the magnitude of the largest weight is a good (though not perfect; see page 120) measure of how close we are to a local minimum of the negative log likelihood criterion being minimized. If the specified number of epochs passes without this ratio beating its minimum so far, convergence is said to have been achieved.

Max epochs: This is a backstop, in other words, insurance against endless iteration. It should never be used as an actual convergence criterion because it is a brute-force rule, with no intelligence about actual convergence. Make it large, and trust that except in very rare pathological situations, one of the main convergence criteria will handle the situation well.

Visible mean field (vs stochastic): If this box is checked, the reconstruction of the visible layer will use the mean field approximation of Equation 3-4. If not checked, the reconstruction will sample according to Equation 3-2. It is likely that using the mean field approximation is best, although this is not universally agreed upon. In practice, the difference seems slight.

Greedy mean field: If this box is checked, propagation of input data through early layers for greedy training strictly uses mean field approximations. If not checked, sampling is done for the inputs to the layer being trained (except the first layer, which is never sampled). This topic is discussed in detail on page 173.

Binary splits: If this box is checked, the raw input data will be quantized to strictly binary data by setting variables above their mean to one and those equal or below the mean to zero. If not checked, the raw input data will be linearly scaled to a range of 0–1.

Fine tune complete model: If this box is checked, after the entire deep belief net is constructed (all RBMs greedily trained, then all subsequent layers trained with supervision), supervised training will be used to tweak the entire model, including the RBM layers. This will always improve in-sample performance and often improve out-of-sample performance. But display of reconstruction samples becomes pointless garbage.

Supervised Training Params

This menu option sets the parameters that are relevant to supervised training of the layers following the RBM layers, as well as the optional fine-tuning of the complete deep belief net. All parameters are preset to defaults that should be reasonable for many or most applications. The following parameters may be set:

Subsets to prevent CUDA timeout: This has no effect whatsoever on the model produced. It affects only the degree to which computations are split up; the results of the computations remain the same. *This is different from batches in RBM training.* RBM batch division does impact the model and the nature of convergence because the weights are updated for each batch. In supervised training, all batches are pooled, with one weight update per epoch (pass through the entire training set). To reduce kernel-launch overhead, the number of subsets should be set as low as possible. But keep an eye on the CUDA time summary in CUDA.LOG and be prepared to use more subsets if any time-per-kernel approaches the two-second Windows limit.

Annealing iterations for supervised: This is the number of simulated annealing passes used to find a good weight set from which to begin training. This topic is discussed in detail in the section that begins on page 78. This is usually a fairly cheap operation with good returns for the first few hundred passes. More than a few thousand iterations is probably overkill because of rapidly diminishing returns.

Initial random range: This is the average range of weight perturbation for simulated annealing. The program will periodically raise and lower the user-specified figure to make this parameter less critical. For this reason, the progress plot of error will have clearly visible periodic variation. This is normal operation. The exact algorithms that govern simulated annealing perturbation are shown in the section that begins on page 78.

Supervised max iterations: After RBM training is complete, the supervised layers following the RBM layers are trained. This parameter limits the number of epochs to prevent wildly excessive runtimes. It should be set to a very large value and used as insurance only, not as the usual convergence determiner.

Supervised convergence tolerance: This is the primary method for determining convergence of training the supervised layers. Training is stopped when the relative change in the error from one epoch to the next falls below this level. Because the supervised training algorithm used in DEEP is deterministic, this can safely be set to a very small value, although doing so is usually without merit because most improvement happens early in training.

Complete max iterations: This is identical to "Supervised max iterations" except that it applies to the optional fine-tuning of the complete (unsupervised RBMs plus subsequent supervised layers) deep belief net.

Complete convergence tolerance: This is identical to "Supervised convergence tolerance" except that it applies to the optional fine-tuning of the complete (unsupervised RBMs plus subsequent supervised layers) deep belief net.

Weight penalty: This penalty discourages large weights during supervised training. It should nearly always be set to a very small value, small enough that it does not have an overly strong impact on learning "best" weights yet large enough that it prevents the large weights that can happen in some unusual pathological situations that are especially likely when the inputs to the supervised section are strongly correlated. This topic is discussed in detail beginning on page 27.

Is model a classifier: By default, MNIST data produces a classifier model, and database data produces a predictive model. This option allows the user to override the default. If database data is read and the user forces the model to be a classifier, at least two targets must be selected, and for each case the target having greatest value is assumed to be the correct case.

Prohibit singular value decomposition: The section beginning on page 80 discusses how the extremely efficient singular value decomposition (SVD) algorithm can be used to explicitly compute optimal output weights for a predictive model and discover excellent starting weights for iterative training of classifiers. But for gigantic problems and in some rare pathological situations, SVD can fail, even (rarely) producing not-a-number results. For this reason, SVD is disabled if there are more than 400 inputs to the output layer. Moreover, the user may choose to disable SVD. Because SVD is such an enormous help in achieving rapid and high-quality convergence, it should always be allowed if at all possible.

Train

The Train selection trains the entire deep belief net. First, all RBM layers are trained with unsupervised greedy training. Then, all subsequent layers (typically just one, the output) are trained using supervision. Finally and optionally, the entire deep belief net is fine-tuned with supervision. The steps for complete training are shown on the left side of the screen. Those that will not be used in the current configuration are grayed out. A marker arrow identifies the step currently executing, and particularly slow operations indicate the percent completion.

The first step in RBM training is to find initial weights by randomly generating weight sets and finding the one with a minimum reconstruction error. In Figure 5-1, below we can see this operation in progress. The top line on the left side says that we are training RBM layer 1. The initial weight operation is 55 percent complete. The graph is the RMS reconstruction error, with the light blue line showing the individual tries and the heavy black line showing the best so far.

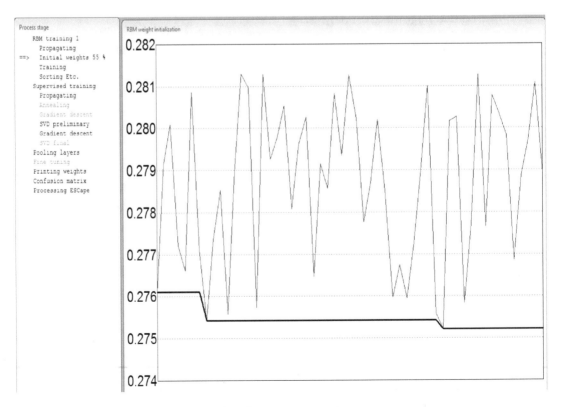

Figure 5-1. *Finding initial weights for RBM training*

After initial weight selection is complete, the program trains the RBM using stochastic gradient descent. The screen will resemble Figure 5-2.

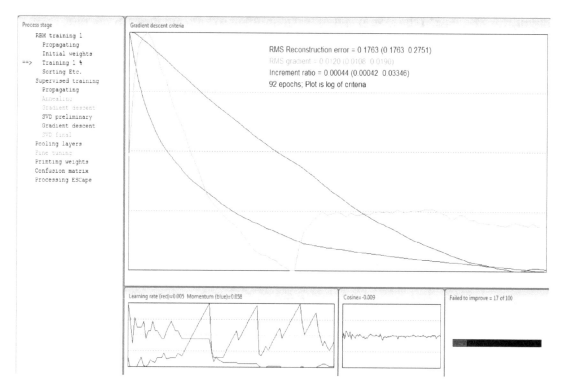

Figure 5-2. *RBM training*

On the left side we can see that we are in the Training operation and we are 1 percent done. This percentage is relative to the "Max epochs" parameter, which, as stated earlier, should always be set overly large and used only as a backstop. Hence, this percentage will nearly always be very pessimistic relative to the actual training progress.

The largest window plots three values, whose current, minimum, and maximum values are written in the top center of the plot. The reconstruction error is in red, and it typically drops off fast and then levels out. The increment ratio (the maximum increment divided by the maximum weight) typically decreases fairly linearly before hitting a sharp knee and flattening, with a few subsequent small bounces. The RMS gradient often displays peculiar behavior, with very gradual decreases punctuated by sharp jumps up as blocks of weights suddenly go from near zero to larger, more useful values.

Be aware that these plots are the logs of the values, not the actual values. Also, each plot is scaled so that the entire historical range of the parameter exactly covers the vertical extent of the plot. The net effect is that as training progresses and values become small, tiny changes in the actual values are magnified to large changes in the plot. This magnification is useful in that it shows in great detail exactly what is happening. Unfortunately, it can be deceptive, making the user think that violent gyrations are occurring when, in fact, the changes in the actual values are miniscule.

The lower-left graph shows the dynamically adjusted learning rate and momentum, also scaled so that the historical values exactly fill the vertical extent of the plot. Typically, the learning rate will show a net decrease, dropping to a very small value after several dozen iterations during which it bounces. The momentum only rarely stabilizes, climbing steadily until it becomes excessive and causes an overshoot that results in backtracking, at which point the adjustment algorithm slaps it back down for a while.

The bottom-center graph shows the cosine of the angle between successive gradients, scaled to a fixed range of minus one to one. It should always be near the center, indicating that the weight increments are neither undershooting nor overshooting.

The lower-right bar graph shows the number of contiguous failures of the increment ratio to decrease, relative to the user-specified limit. When the red interior reaches the right side of the bar's outline, training will terminate. This is the primary convergence criterion.

Supervised training of the post-RBM layers, as well as the optional fine-tuning, also cause graphs of the error to be displayed as training progresses. There is nothing fancy or confusing about them, so we'll dispense with a detailed discussion.

Test

The Test selection tests the trained model on the current dataset. There is little point in training and then immediately testing a model because the test would just reproduce the same results given when training is complete. However, this selection facilitates testing the model on new data.

The usual procedure for training and testing a model is as follows:

1. Read the training data.

2. Define the architecture.

3. Select the predictor and target variables.

4. Set training parameters if something other than the default is desired.

5. Train.

6. Clear all data.

7. Read the test data.

8. Test.

The test dataset must contain the same variables in the same order as the training dataset. The user must not change the architecture or the predictor/target variables.

Note   The Test option does not use CUDA processing. If the model was trained with CUDA enhancement, it is possible that the slightly different floating-point computations with and without CUDA may result in slightly different test results. Any differences should be small.

Analyze

This selection computes and prints to the DEEP.LOG file two tables of information. The first is a comparison, for each input variable, of the probability of it being activated in the training set versus the probability of it being activated in the reconstructed input layer. Here is a short segment illustrating this table:

Variable	Visible	Reconstructed
P_8_10	0.616	0.617
P_8_11	0.551	0.547
P_8_12	0.522	0.519
P_8_13	0.516	0.513
P_8_14	0.517	0.511
P_8_15	0.520	0.514
P_8_16	0.517	0.513
P_8_17	0.514	0.510
P_8_18	0.539	0.536
P_8_19	0.606	0.603
P_8_20	0.706	0.706
P_8_21	0.806	0.810
P_8_22	0.887	0.891
P_8_23	0.942	0.943

The other analysis output is the probability (across the training set) of each final (topmost) layer hidden neuron being activated. Here is an example of this table:

Hidden	Activation
1	0.837
2	0.449
3	0.723
4	0.596
5	0.578
6	0.501
7	0.501
8	0.418

Receptive Field

The *receptive field* of a hidden neuron in an RBM is (loosely) defined as the pattern of weights connecting the input layer to the hidden neuron. If the input happens to be an image, such as is the case with MNIST data, then it is possible to display these weights in the same dimensions as the input image. Figure 5-3 shows the receptive fields of a dozen neurons trained with MNIST data. Large positive weights are white, large negative weights are black, and intermediate values are shades of gray. A color display is also an option, with positive weights colored cyan, negative weights colored red, and brightness corresponding to magnitude. The gray areas around the perimeter are pixels that are constant for all cases and hence omitted from the model.

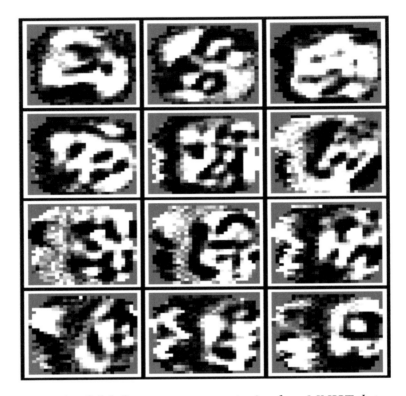

Figure 5-3. *Receptive fields for some neurons trained on MNIST data*

Generative Sample

We saw on page 175 that a trained RBM or set of RBM layers can be made to spit out random samples from the distribution that it has learned. Examination of such random samples can be interesting because they show examples of the primitive patterns that the model has learned.

As with a receptive field display, this option is valid only for MNIST images (at least in DEEP version 1.0). Again, the user must specify the number of rows and columns to display. Each of the *nrows*ncolumns* images is a separate sample.

As discussed on page 175, there are two ways to begin the Markov chain whose final value will be the computed sample. One can begin with a member of the training set. To do this, set the "First case" field to a positive number, the sequential number of the training case that will be used for the first sample. Subsequent samples will start from subsequent training cases. The degree to which the final reconstruction resembles the starting pattern is an indication of the quality of training and the degree to which efficient mixing is taking place in the Markov chain.

Figure 5-4 shows the first 12 cases from the MNIST test set of 10,000 cases. Figure 5-5 shows generative samples obtained from these cases using 10,000 iterations. What makes this interesting is that this was derived from a single RBM layer having just 15 hidden neurons! The degree to which this tiny model has encapsulated training set patterns is astounding.

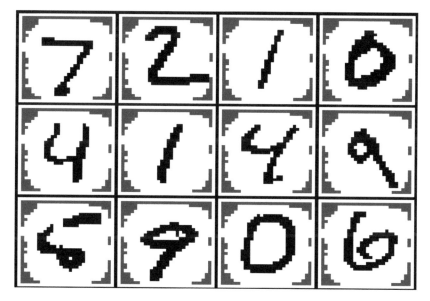

Figure 5-4. *First 12 cases of MNIST test set*

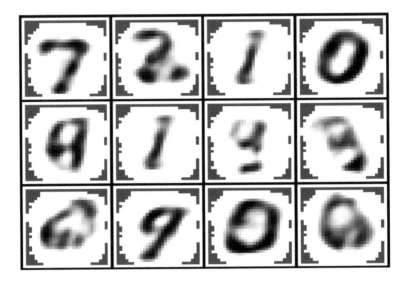

Figure 5-5. *Generative samples after 10,000 iterations*

Alternatively, one can set the topmost hidden neuron layer to random values, thus divorcing the computed samples from training data. This lets us see the actual primitive patterns that the model is recognizing. Figure 5-6 shows 108 random samples obtained from an RBM having 100 hidden neurons, using 50,000 iterations. Note that if one were to append ten more visible neurons to the input layer to identify the digit represented, an option that may be added in a future version, and then clamp these ten neurons to the "correct" value, one would see representations of the actual digits. This is not done here, so rather than seeing *digits*, we see the *components* of the digit images that the model has learned.

Figure 5-6. *Samples using 100 hidden neurons randomly set*

The DEEP.LOG File

When a database or MNIST file is read, the program creates a new file called DEEP.LOG in the same directory as the data file being read. If a file of that name already exists, it is erased. This log file begins by showing the directory in which it is created, along with the date and time. It then lists the mean and standard deviation of every variable read. Here is a typical example:

Deep (D:\DEEP\TEST\DEEP.LOG) 1/26/15 15:42:16

Found 23 variables in input file D:\DEEP\TEST\SYNTH.TXT

6304 cases read

Means and standard deviations...

Variable	Mean	StdDev
RAND0	0.00711	0.57541
RAND1	0.01422	0.58043
RAND2	0.01027	0.57694
RAND3	-0.00765	0.58143
RAND4	0.00713	0.57911
RAND5	-0.01166	0.57263
RAND6	-0.00648	0.57742
RAND7	-0.01424	0.58015
RAND8	0.00659	0.57533
RAND9	-0.00366	0.57733

It then shows the architecture of the model, including the unsupervised and supervised sections.

Beginning training a model with the following architecture:

There are 1 unsupervised layers, not including input
  Hidden layer has 5 neurons

There are 1 supervised layers, including output

Since there is at least one RBM layer, the training parameters for this layer are listed.

Restricted Boltzmann Machine training parameters...
  Initial random iterations for starting weights = 50
  Number of batches = 24
  Markov chain length start = 1
  Markov chain length end = 4
  Markov chain length rate = 0.0050
  Learning rate = 0.05000
  Starting momentum = 0.10000
  Ending momentum = 0.90000
  Weight penalty = 0.00010
  Sparsity penalty = 0.00100
  Sparsity target = 0.10000
  Increment convergence criterion = 0.00001
  Max epochs with no improvement = 500
  Max epochs = 10000
  Visible layer using mean field, not stochastic
  Inputs will be rescaled to cover a range of 0-1
  Unsupervised section weights will be fine tuned by supervised training

The training parameters for the supervised section are also listed.

Supervised layer(s) training parameters...
  Initial annealing iterations for starting weights = 100
  Initial random range for starting weights = 1.00000
  Supervised optimization max iterations = 1000
  Supervised optimization convergence tolerance = 0.0000500
  Complete model optimization max iterations = 2000
  Complete model optimization convergence tolerance = 0.0000100
  Weight penalty = 0.00100

The results of training the unsupervised layer are printed first.

Training unsupervised layer 1
  Initial weight search RMS reconstruction error = 0.27098
  Unsupervised training complete; RMS reconstruction error = 0.31654

There is one curious issue in that result. The initial weight search gave a reproduction error of 0.27098, but after real training was done, we see that the reproduction error has increased to 0.31654. How did this happen?

Actually, this is unusual, happening only when the input variables have little or no patterns that the RBM can learn. In this example, the inputs are all random numbers, so there are obviously no patterns. We must remember that the reconstruction error is measured slightly differently during weight initialization and training. In the section that begins on page 93, we see that the initial search reconstruction error is computed in a deterministic manner using mean field approximation in both directions. But during learning we use random sampling of the hidden neuron activations for the reconstruction error, as discussed on page 100. This tends to increase the error somewhat. If the RBM is able to learn real patterns, the difference because of randomization during the reconstruction error computation is swamped out by the model's ability to reconstruct authentic patterns. But if there are no patterns to reconstruct, we just get the effect of randomization.

After the greedy training of the unsupervised section is complete, the supervised section that follows the unsupervised section is trained. Fine-tuning was selected, so the last step is to tweak the entire model, unsupervised plus supervised sections. Here we see that fine-tuning produces a huge improvement in the criterion, which is negative log likelihood in this example because a classification model was forced.

Optimization of supervised section is complete with negative log likelihood = 0.12270
Fine tuning of the entire model is complete with negative log likelihood = 0.02327

The targets are listed, and it is noted that the inputs are rescaled 0–1, so the weights that will be printed soon refer to these rescaled values.

Trained weights for this model, predicting the following target(s)...
  RAND1
  RAND2
  RAND3

Each raw input has been rescaled 0-1 to cover the min/max range.
Thus, all weights refer to the rescaled value, not the raw value.

The weights for the single unsupervised layer are now printed. If there were multiple layers, each set of weights would appear. These weights are *after* fine-tuning.

Weights for unsupervised hidden layer 1

	1	2	3	4	5
Q mean	0.4522	0.4796	0.4556	0.4138	0.4717
skewness	0.1310	0.0700	0.1271	0.2440	0.0618
RAND1	-7.0347	-4.5028	0.9392	-2.5469	1.4879
RAND2	4.7047	-1.7225	-0.5104	7.0462	2.0824
RAND3	2.8726	6.0903	1.6480	-4.7952	-2.6467
RAND4	-0.0131	0.1551	-1.6304	-0.2535	0.3858
RAND5	-0.0032	-0.3523	-0.0453	0.1271	-0.8947
RAND6	-0.0619	-0.1453	-1.8291	-0.1889	-0.2881
BIAS	0.7231	-0.5790	0.8983	-0.6242	0.4237

The model was specified to have five hidden neurons, so we have five columns, one for each. At most ten columns are printed. After each unsupervised layer is trained, the hidden neuron weights are sorted so that the hidden neuron having maximum sum of absolute values becomes the first hidden neuron, and so forth. This way, if we examine the weights to obtain hints about the interpretation of features detected, we can focus your efforts on the early columns. However, if fine-tuning is done, as is the case in this example, this sorting can be subverted. This is not a practical problem because fine-tuning almost always largely or entirely destroys the interpretability of weight patterns that were discovered by the RBM.

The "Q mean" row is the mean activation of each hidden neuron, and the "skewness" row is the statistical skewness of the activations. In general, a positive skewness means that the neuron is usually off, and vice versa. These two values are computed *before* fine-tuning; they refer to the actions of the trained RBM before its weights are adjusted by supervised fine-tuning.

We then see the weights that connect the (last and only) unsupervised layer to the (first and only) supervised layer. Also, the final value of the optimization criterion, which we saw earlier, is repeated.

Weights for final (output) layer

Target 1 of 3: RAND1

   -9.158017 Unsupervised output 1
   -6.844571 Unsupervised output 2

    0.781757 Unsupervised output 3

    -2.436789 Unsupervised output 4

    2.660202 Unsupervised output 5

    6.449515 CONSTANT

Target 2 of 3: RAND2

    5.160418 Unsupervised output 1

    -1.469535 Unsupervised output 2

    -0.629605 Unsupervised output 3

    9.063721 Unsupervised output 4

    2.708100 Unsupervised output 5

    -8.018184 CONSTANT

Target 3 of 3: RAND3

    3.467198 Unsupervised output 1

    8.798016 Unsupervised output 2

    1.170767 Unsupervised output 3

    -8.699801 Unsupervised output 4

    -3.433103 Unsupervised output 5

    -2.159271 CONSTANT

Negative log likelihood = 0.02327

Lastly, the confusion matrix is shown. Usually, when one is training a classifier, the target vector for each case has 1.0 in the position corresponding to the correct class and 0.0 in all other positions. But this is just a common convention and is not required in DEEP. Instead, whichever target has the maximum value is defined to be the correct class. So, when a model having continuous targets is forced to be a classifier, as is the situation in this example, the results are reasonable. In particular, we would expect good classification in this example since all three targets are also present as inputs! Indeed, we can see this is the case.

Confusion matrix... Row is true class, column is predicted class
  In each set of three rows for a true class, the first row is the count,
  the second row is the percent for that row (true class)
  and the third row is the percent of the entire dataset.

	1	2	3
1	2128	3	8
	99.49	0.14	0.37
	33.76	0.05	0.13
2	9	2088	15
	0.43	98.86	0.71
	0.14	33.12	0.24
3	8	7	2038
	0.39	0.34	99.27
	0.13	0.11	32.33

Total misclassification = 0.7931 percent

Index

A

Activation file, 184, 187
Architecture of CUDA, 37–38

B

Backpropagation of errors
    chain rule, 10, 11
    derivative of, 12
    hidden-layers, 11, 12
    hidden neurons, 11
    mean squared error, 9–10
    MLFN, 10
    neurons activation, 10–11
    parameters, 9
    SoftMax outputs, 13–16, 18

C

Clear all data option, 184, 187
Conjugate gradient optimization
    change, 86
    concept of, 86
    difficulties, 85
    dimensions, 87
    function minimization, 86
    implementations, 84
    multivariate, 89
    negative gradient, 89
    power of, 88
    second-order methods, 85
    simple quadratic function, 88
    traditional backpropagation, momentum, 89
    vectors, 86
CUDA code, RBM training
    bias vectors and weight matrix, 157
    calling parameter list, 153
    components, 153
    development, 172
    device and clean up, parameters, 159
    epoch loop, 155
    fetching, 143–144
    gradient length and dot product, reduction, 146
    hidden-to-visible analysis, 145
        issue efficiency and stall reasons, 170
        load balancing, 170
        memory access statistics, 171
        occupancy, 169
        pipe utilization, 171
    initialization and cache line matching, 142
    local variable declarations, 154
    maximum weight, 159
    memory allocation, 155
    random training data, 160
    ratio criterion, 159
    reconstruction error, 156–157
    thresholds and adjustment factors, 158

© Timothy Masters 2018
T. Masters, *Deep Belief Nets in C++ and CUDA C: Volume 1*, https://doi.org/10.1007/978-1-4842-3591-1

H, I, J, K, L

Get the eBook for only $5!

Why limit yourself?

With most of our titles available in both PDF and ePUB format, you can access your content wherever and however you wish—on your PC, phone, tablet, or reader.

Since you've purchased this print book, we are happy to offer you the eBook for just $5.

To learn more, go to http://www.apress.com/companion or contact support@apress.com.

Apress®

Printed in the United States
By Bookmasters